THE SECRET OF GRAVITY
AND
OTHER MYSTERIES
OF THE
UNIVERSE

This book contains:

248 pages

1,360 paragraphs

3,859 sentences

9.156 lines

66,613 words

331,209 characters

with

17 average words per sentence

67 maximum words per sentence

average word length = 5

and

51 illustrations

COVER GRAPHICS

The graphics illustration on the front cover was created from a mathematical formula using the Mathcad computer simulation program (copyright © 1991-1997 MathSoft, Inc.). The graph pictures the distortion of a rotating field that is produced when the circular velocity is delayed as a function of the length of the radius. One axis of the field is seen to wrap around the center of the field, forming the shape of a spiral. A distortion of this type can be formed by a radial time delay as the field rotates. A delay of this type may be attributed to the *ether* of the universe, if the universe is, indeed, permeated by an ether. The great scientists of the past were intrigued by the possibility that an ether is present in the universe, while contemporary scientists have less faith in its existence. The analysis of Chapter VIII provides enough substance to entertain reconsideration of the presence of an ether.

THE SECRET OF GRAVITY AND OTHER MYSTERIES OF THE UNIVERSE

Weldon Vlasak

Adaptive Enterprises
Clatonia, Nebraska

THE SECRET OF GRAVITY AND OTHER MYSTERIES OF THE UNIVERSE

Published in 1997 by :
                      Adaptive Enterprises
                      RR #1, Box 73
                      Clatonia, NE 68328

Copyright © 1997 by Weldon Vlasak.

All rights reserved. No part of this book may be reproduced or transmitted in any form or by any means, electronic or mechanical, including photocopying, recording, digital communications, or by any information storage and retrieval system without written permission from the author, except for the brief inclusion of a quotation in a review.

First edition 1997
First printing 1997

Library of Congress Catalog Card Number 90-094377

ISBN 0-9659176-0-6

Dedicated to:

*My wife, Marge,
who made many sacrifices in
the production of this book
and has supported me
throughout our lifetime together.*

*And also to my children,
Jackie and Rick,
with whom we
shared life's experiences
during their formative years.*

ACKNOWLEDGEMENTS

Any accomplishment necessarily follows from the efforts of those who have previously contributed to the foundation of human knowledge. Earlier scientific achievements date back over a thousand years, and Chapter I is dedicated to those who have made outstanding achievements. Any challenges to existing theories that are presented in this book does not diminish my respect for these scientists, whom I hold in the highest regard.

Whatever engineering capability that I have attained must be attributed to my educational background period, during which many college professors contributed to the degree of my comprehension of difficult subjects. I do not remember all of their names any longer, but certainly Dean of Engineering, Thomas L. Martin, (formerly of the University of Arizona), Dr. Curtis, Professor of Mathematics (formerly of the University of Miami), and Professors Meltzer, de Pian, Wheeler and Kyriakopoulos (who were members of the staff of The George Washington University) stand out in my mind. My apologies to all of the others whom I have unintentionally neglected.

Dr. Frank Reuter is highly commended for his numerous useful suggestions and corrections, in spite of the difficulties that were, by necessity, imposed upon him.

And thanks to Wood Sabold who made numerous useful suggestions.

TABLE OF CONTENTS

INTRODUCTION   15

CHAPTER I   19
SCIENCE OF THE PAST
    Surprises at the end of the 19th century, new problems, new answers.

CHAPTER II   31
DIFFICULTIES WITH PRESENT PHYSICAL THEORIES
    Conflicting theories, strengths and weaknesses of quantum physics, and visions of reality.

CHAPTER III   39
OUR CIRCULAR UNIVERSE & RUTHERFORD'S ATOM
    Why the universe is circular in nature.

CHAPTER IV   49
THE "BIG BANG" THEORY
    The universe viewed as an expanding bubble, its rate of expansion, and the age of the universe.

CHAPTER V   67
INNER SPACE
    The construction of atoms: a mechanical model.

CHAPTER VI   81
A UNIVERSAL FORCE
    The present scientific view of the forces of the universe, and the dilemmas produced by the premise of a single universal force.

CHAPTER VII   93
RESOLVING THE GRAVITATIONAL FORCE DILEMMA
    A new approach, and the clue to solving the greatest mystery of all time.

CHAPTER VIII   97

THE ROTATING DIPOLE OF THE HYDROGEN ATOM
Analysis of the rotating dipole of the hydrogen atom.

CHAPTER IX   111
THE GRAVITY EQUATIONS
The solution to the mystery of the cause of the gravitational force.

CHAPTER X   127
ELECTRICAL CONDUCTION AND MAGNETISM
What produces electrical currents, and a new view of magnetism.

CHAPTER XI   139
A UNIVERSAL FIELD THEORY
A single type of field accounts for all of the dynamics of the universe.

CHAPTER XII   145
THE PROBLEMS WITH GRAVITY
Previously known problems with gravitational attraction, and some new ones.

CHAPTER XIII   155
WHAT PRODUCES ATOMIC SPECTRA?
A classical physical analysis of how the discrete spectral lines of emission are produced by excited atoms.

CHAPTER XIV   163
ATOMIC ORBITALS
The current calculated shapes of the orbitals of the electrons in atoms are replaced by more realistic forms.

CHAPTER XV   179
THE SPEED OF THE ELECTRON IN ORBIT
    The velocity electrons moving in atomic orbits is estimated by a different method.

CHAPTER XVI   183
SPEEDS EXCEEDING THE SPEED OF LIGHT
    Einstein's conclusion contested.

CHAPTER XVII   191
IS THE ELECTRON A WAVE?
    A basic dilemma of physics is examined.

CHAPTER XVIII   195
THE ELECTRICITY WITHIN US
    An estimate of the electricity flowing within our bodies.

CHAPTER XIX   197
PUTTING IT ALL TOGETHER
    The information in the previous chapters is woven together to present a new and different picture of what is happening around us.

CHAPTER XX   213
AFTERWARDS
    Where do we go from here?

APPENDICES   217

INDEX   243

Preface to the First Edition

The secret of gravity has finally been solved. What at first seemed so difficult now appears simple. How could so many scientists miss it? It seems incredible. But this book is not just about the forces of gravity. The use of unconventional methods in the investigation leads toward the solutions of other various mysteries of science.

This book is intended for those who have an open mind, those who are interested in science, and those who want to learn new ideas and listen to new thoughts. Some of the most difficult problems of science have been assaulted, and the result is a new and different view of the universe. A scientific background will definitely be helpful to the reader. Hopefully, those with less scientific knowledge will also find something useful in this book. The first part of the book is written in very simple terms, and the young student may find it useful in getting a perspective view of science

Different approaches have been used in the investigation, leaning more toward engineering methods than those of physics. The resulting new and different conclusions may prove to be disturbing, since they do not agree with present theory. It is not easy to change one's point of view. It was also not easy for me to publish material that may turn out to be contradictory.

When a subject of this type is investigated, the path that is followed is not always one of choice once the original direction is chosen. So, while the general subject matter is woven together as well as possible, the plot is not the same as a novel, and discoveries flow with the waves of chance. All chapters, are, however, related to one another.

I have tried to make the material as simple as possible, but some of the concepts are highly complex and difficult to explain. Therefore it was necessary to include some mathematical equations in the body of the text. The more complex mathematics of the Appendix are considered to be important proof of the secret of gravity, although not everyone is expected to have a mathematical background and be able to follow the proofs. In any case, I hope that you will find this book interesting, challenging, and certainly *different*.

Imagination rules the world.

---**Napoleon**

The only thing certain is that nothing is certain.

---Pliny the Elder

Introduction

What you are about to discover may sound like science fiction. I can assure you that it is not. While imagination has played its part, the scientific investigations upon which this book are based are on a solid science and engineering methods. Many hours of difficult research were spent on the studies, which took place over a period of several years.

The time has now come to unravel the secret of gravity. While some will find the result surprising, certain theoretical physicists may have suspected the answer that is forthcoming, at least to some degree. The puzzle of gravity has long been regarded as the final hurdle to be overcome in developing a unified field theory, something that Einstein sought but was unsuccessful in accomplishing. Indeed, the road leads in that direction, and the foundation of an all-encompassing theory that pictures a universe that is filled with powerful fields is established. Ours is a universe that is both chaotic and unbelievably stable; a universe of large numbers but, as far as can be determined, finite.

In the forthcoming chapters, some of the most challenging mysteries of the universe are examined. These investigations cover space from the tiniest dimension to the outer limits of the universe. While neither of these limits, inner or outer, can be observed directly, sufficient evidence exists to provide the necessary clues to some scientific mysteries that have never before been solved. These clues are used to good advantage. It is not possible, however, to solve any mystery of great depth and complexity without adopting new and different approaches and abandoning or modifying accepted premises. That is why unconventional ideas and methods have been employed.

The book begins by examining specific turning points in scientific history up to this time. The known information about the subject is examined, and then conclusions are formed by a reasoning process and suppositions are formulated. The suppositions are exercised in various ways, and assertions are made. Finally, the results are tested and data is collected and illustrated. These basic logical methods are not new, dating back to the time of the ancient Greeks such as Socrates, Plato and Aristotle.

Many of the books that have been written on the subject of physics and cosmology tend to be in some form of historical revelation or recount. This choice of treatment is probably due to the complexity of the subject matter, and also because many great scientists have already made thorough investigations and have made countless measurements, which makes any further contribution seem unlikely. While it is necessary to include some historical background information to show what has already been accomplished, this book is not a mere historical account of science. Several new and bold ideas will be presented. The book begins by noting the great accomplishments of the past that apply to the subject matter of following chapters.

Cosmologists have devised theories that explain how the universe may have been created. One of these, the "Big Bang" theory, is well-accepted, and many scientists have contributed to the model which describes the physics of energy, time, and dimension that describe the formulation of the structure of the universe. While the Big Bang theory is generally accepted within the scientific community, it is also true that many doubts exist about many of the assumptions that are necessary to cover the expanse of time from the first instant of its creation to the present time. The Big Bang theory will be examined, and the more troublesome suppositions of cosmology will also be discussed.

Some may find it difficult to accept many of the assertions that are necessary to support much of theoretical physics. The scientific methods of *quantum physics*, a major current accepted methodology of physics, has proven to be quite successful but is not without its weak points. Quantum physics is based on the fundamental principles of conservation of energy, and some of the more important areas of this field will come under discussion. The chosen methods utilized in this book are based primarily on *force*, as was Newton's Theory of Gravity.

A new theory of the phenomena that produces the gravitational force is proposed in a later chapter. It is surprising that not many theories could be found which provide any explanation of the cause that produces the force of gravity. Certain popular theories are seldom confronted or even questioned, such as Einstein's Theory of Relativity, electrical concepts, the speed of light, and Maxwell's Theory of Radiation. Several of these theories will, however, be challenged, as must be the case for any thorough investigation. New approaches have led to new ideas on these subjects, and some new questions and new dilemmas are revealed and will also be discussed.

Once the problems with current scientific theories are outlined, the direction of the investigation is defined. The remaining chapters present some new and different thoughts and methods of analysis. I believe that, as you read through these chapters, you will see that some weaknesses and lack of rational thought still exist in current scientific theory. The attempt has been made to use engineering methods to establish better and clearer visions of our universe and everything in it.

Beliefs continually change, to some degree, with time. In seeking a new vision of the truth, we will begin with questions. The main question is "What is the secret of gravity?" This leads to other questions, such as:

"How big is the universe?"
"What is magnetism?"
"What is electrical conduction?"
"Is it possible to exceed the speed of light?"
"How is the spectrum of radiation produced"
"What produces the gravitational force?"
"Is there a universal *field*?"

These are very deep questions that have been asked many times by many people, and yet the answers are still incomplete, even after extensive investigations performed by numerous highly qualified scientists. To get the right answers, one must ask the right questions. It is hoped that you will find the answers that will be revealed in the following chapters are at least interesting and that you will acquire a new vision of the universe as a result.

The solution to the secret of gravity unlocks the door to a unified field theory. It is fortunate that so much data about the universe is available as result of so many efforts by so many brilliant scientists.

Without a rich data bank, the above problems probably could not be solved by any one person, since the degree of effort would be beyond any individual effort. Even with the proper tools, the task is formidable, and this writer had very limited resources.

In this book, very little emphasis has been placed on the utilization of the research efforts of the past five years. The reason is that recent efforts have been concentrated on high-energy particle research and the exploration of outer space, and these subjects tend to diverge from the subject of this book (the work of Hubble, in measuring radiation from outer space, are, however, useful).

Not all of the mathematical formulas that were employed in the investigations (that were made in the process of developing the background material for this book) have been included. Descriptions are provided, wherever possible, in layman's terms for ease of reading. Simplifications, however, are not always possible, and some use of mathematics is necessary. In particular, the gravitational equations are rather complex and are placed in the Appendix. These equations provide important evidence that is essential to the arguments that are made. Readers with a background in math or science should be able to interpret most of the material provided, while others may not want to look at equations. The curious may want to see the results of the investigation and how the conclusions were reached.

While some of the most important mysteries of science are believed to be solved, other curious dilemmas show up, and some of the investigations could not be completed, simply due to a lack of resources and time. In spite of the shortcomings, the effort has a payoff, and it is hoped that this book will stimulate the reader's interest in science and provide a new vision of the universe.

Will the road of advance again make a sharp turn, as it has so often done in the past?

---A. Einstein

What's past is prologue.

---Shakespeare

CHAPTER I

Science of The Past
Many Problems and Some Surprises

The following account is an overview of some of the important developments in science that apply to the subject of this book. Since this is neither a historical book nor a text book, only selected major events are described. References are given in the bibliography for those who wish to obtain greater details.

One of the subjects of this book is electricity. While most of the developments in the field of electricity have been comparatively recent, the existence of electric phenomena was recognized long ago. Static electricity is believed to have been discovered in the 6th century B.C. Two materials, when rubbed together, produce small electric arcs. We now believe that static electricity is a function of the characteristics of atoms and molecules of different types of matter, but very little was known about static electricity at the time. It is therefore surprising that atomic theory, which has progressed rapidly within the last two hundred years, was first conceived in the 4th century B.C. by an ancient Greek, Democritus, who originated the idea of the atom as the smallest particle of matter.

Many of the basic reasoning methods of science were originated by the Greeks, who undoubtedly used many of the techniques of mathematics and engineering developed by the ancient Egyptians in their teachings. The ancient Greeks tried to find solutions to the mysteries of nature using philosophical methods. Their approach was to postulate premises (which could be intuitive) and then use logical deduction to derive the consequences obtained from the resulting conclusions. Empedocles, in the 5th century, incorporated the proposition that there are but four elements - earth, water, air, and fire, into his scheme of the universe. Socrates (469-399 B.C.) was a great sage and teacher who devised a philosophical thought process consisting of posing questions and eliciting answers, followed by adroit questions which made the errors in the answers conspicuous (Socratic irony). Plato (437-347 B.C.) was a student of Socrates who became know for his philosophies. In his doctrine, things are copies of ideas, and these ideas are the object of true knowledge. Aristotle (384-322 B.C.), who was taught by Plato, was less concerned with the metaphysical and adopted empirical methods and used *formal logic* to form conclusions. Logical reasoning is at the heart of scientific methods which are still in existence today.

The characteristics of *mass* will be examined in detail in the forthcoming chapters. Aristotle derived what is probably the first known theory about mass. He noted that a steady force is required to maintain the steady motion of an object which is one of the actions of mass when it is not free of friction. While we now know Aristotle's theory to be incomplete, it lasted for nearly two thousand years. It is also surprising that Aristotle, with his instruments of logic, chose to accept Empedocles idea of but four basic elements in the universe.

Early scientists did not have the tools to examine the universe that we have today and thus had to analyze direct human observations. To them, the earth initially seemed flat, the stars and the planets moved about the earth, and the collections of stars that remained in fixed patterns as they moved across the sky were called constellations in the form of familiar earthly figures. The visualization of outer space changed over the years, evolving into more complex concepts as new ideas came into existence. Ptolemy, as early as the second century A.D., pictured a planetary system which had the planets moving in circles about the earth.

Many of the accomplishments of the ancient Greeks survived for centuries, including Aristotle's somewhat aberrant ideas about mass. Then there were dark periods in man's history, when many efforts were related to wars, politics, and religion. The secretive alchemists initiated the development of chemical procedures, and one of their quests was to find the miraculous "philosopher's stone" which would convert baser metals into gold. They had no way of knowing the difficulty of their quest since they did not realize the amount of power required to alter atoms and because their knowledge of the structure of atoms was lacking.

As science progressed, the importance of accurate data became evident, and measurement methods and standards improved. The definition of the "foot" as a standard of measurement was changed from the length of an individual's foot to a standard length of greater accuracy. Other such standards were established and the art of science began in earnest.

The vision of the universe was changing more radically. A major event occurred when the idea that the earth was not the center of the universe emerged. Copernicus (1473-1573), contemplating the movements of planets and stars, visualized a planetary system wherein the earth rotates about its axis. His revolutionary proposal did not provide answers as to "why do stones fall toward earth if it is not the center of the universe?" or "why aren't birds left behind by the rapidly moving earth?" Copernicus' answer to the last question was that the atmosphere was dragged along with it (a gravitational concept). Unfortunately for poor Copernicus, he was denounced for his ideas as a fool and heretic (as was Galileo), and his theories were condemned as being false and opposed to the Holy Scriptures (by Martin Luther).

The vision of the universe had changed from one which was directly observable to that which requires some degree of imagination in order to extend the mind's picture of space. Tycho Brahe (1546-1601) extended the theory of Copernicus by picturing a geocentric system having the sun at the center of the planetary system. Later Kepler (1571-1630), his student and a mathematician, gave substance to his theory by providing a mathematical description of planetary motion (Kepler was given credit for the theory).

The next major event in the development of the theory of gravity did not occur until the Italian Galileo (1564-1642) showed that a light rock falls at the same rate as a heavy one and also introduced the concept of inertia. Galileo died in the year that the Englishman Newton was born. Sir Isaac Newton (1642-1727) made one of the greatest contributions to gravitational theory. He formulated his law of universal gravitation in 1666. Newton built upon the efforts of Kepler and Galileo to form a mathematical description of the action of the force of gravity. His theory holds that the gravitational force is common to the universe, and that it produces the force of attraction that holds bodies to the earth's surface. Newton's theory produced a greater definition of the force of gravity based on the effects that were observed. He is considered the founder of the laws of gravity. His law of motion, *force equals mass times acceleration*, established the relationship between force, mass, space, and time. The theory of Newton failed to predict a slight variation in the orbit of Mercury because the range of observations that he considered did not include the effect of high velocities of moving objects (mathematical formulas are often incomplete due to the limited extent of the theory when it is first presented).

Investigators were beginning to delve further into the mysteries of matter. In the early 18th century the "phlogiston theory" was introduced by a German named Stahl. All combustible materials were presumed to be composed of "phlogiston," plus other materials, and the phlogiston would escape as a flame when burned, leaving the other materials behind in the ash. Neither the phlogiston nor the theory lasted very long.

Then some basic and important discoveries about electricity were made. Although the electron had not been discovered prior to this time, some rather remarkable theories about the nature of light and electricity had been developed by Volta (1745-1827), Ampere (1775-1836), Gauss (1777-1855), and other scientists near the beginning of the 19th century, and many of these still stand today. These scientists defined the basic variables of electricity (voltage, current, and the magnetic field), and the units of measurement were named after them. The concept of elementary electrical charge was first established in 1833 by Michael Faraday (1791-1867) who performed electrochemical experiments on ionized fluids, and he also described the induction of electrical fields by changing magnetic flux (which eventually resulted in the invention of the electrical transformer). The above investigations resulted in a strong basis for the further development of electrical theory.

Perhaps the most outstanding of the theories of that time (or of any time) was devised by the Scottish physicist James Clerk Maxwell (1831-1871). He established the relationship of the electrical field to the magnetic field (a field is the spatial distribution of electrical forces) wherein a changing electrical field can produce a magnetic field. Maxwell's equations led to his prediction of the radiation of energy which was demonstrated in an electrical experiment by Heinrich Hertz (1857-1894) after Maxwell's death. Maxwell's equations predict that electromagnetic radiation should move at the speed of light. Maxwell's equations, although never proven, led to the development of the radio [demonstrated by Marconi (1837-1937)] and are universally accepted.

Just prior to the turn of the century, laboratory experiments produced events that had never before been witnessed. Certain types of matter were found to radiate energy, and no such phenomena had ever before been observed. Using scrupulous experimental chemical methods, the culprit, the element *radium*, was isolated by the Curies. Why was mass radiating energy? Newton's formulas for mass and force did not suggest such a possibility. A manner of accounting for radiation awaited the discovery of the *electron* which occurred shortly thereafter.

The *proton*, a positive electrical particle, had been observed in 1886 by a German physicist E. Goldstein. Protons were formed in positively charged rays in a discharge tube. A model of the atom was developed (at the end of the 19th century) by a famous English scientist named J. J. Thomson (1856-1940) following his discover of the electron, a negative electrical particle. The spatial distribution of electrical charges within matter was not known at that time, and he portrayed the atom as a sphere with positive charges uniformly distributed throughout (the "plum pudding" model). The plum pudding model lasted for about 14 years.

At the turn of the 20th century, scientists were baffled by the results of experiments which were revealing strange new phenomena that were to change scientific theories and methods radically. Up until that point in time, logical inference and rational thought dominated scientific development. The many advances in chemistry resulted in classifications of basic matter into elements, a process which proceeded into the 20th century. But now matter and its associated mass were not behaving properly, exhibiting strange properties that appeared to be in conflict with numerous prior theories.

Investigations that provided insight as to what constitutes mass (and energy) came about indirectly. Some 100 years ago, a scientist name Wien (1864-1928) was studying the characteristics of energy emitted from a heated sample of material and was able to derive a formula for distribution of energy as a function of optical wavelengths at various temperatures (visible light plus infrared or heat, and ultraviolet). We know, for instance, that iron will turn red if we heat it with a welding torch, and that the color changes from red toward blue with more heat, also becoming brighter. If the surface of the material being heated is black (at room temperature), then the heat radiation characteristics are consistent since the color of the surface affects the amount of radiation, and the energy that is emitted is called "black body radiation."

Instruments had come into being which allowed accurate radiation measurements, and Wien derived a mathematical formula that fitted the radiation curve of black bodies (a model). It was found that the radiation extends over a wide range in wavelengths, and that the visible wavelengths lie in a narrow band at the center of the radiation curve (Figure I-1).

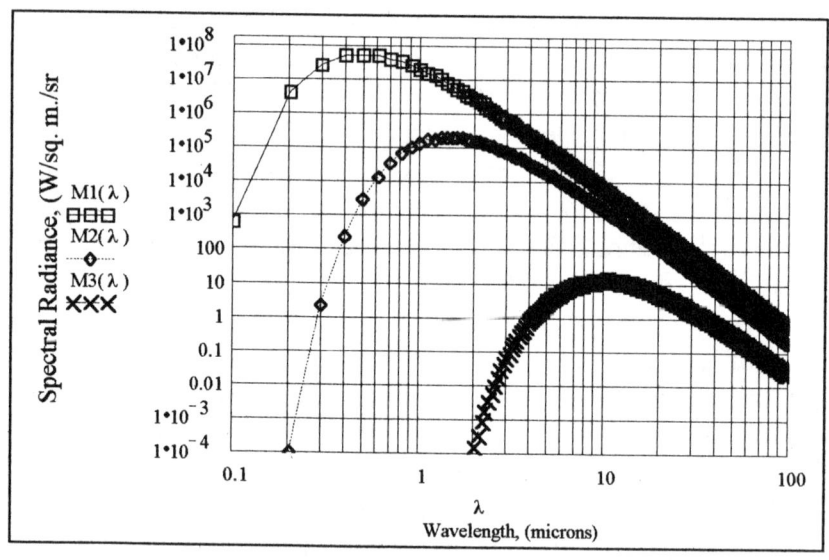

M1 = Radiance at 6000 deg. K
M2 = Radiance at 2000 deg. K
M3 = Radiance at 300 deg. K

Figure I-1. Planck's Radiation Curve
At Three Temperatures

The visible radiation in these curves takes up only a small portion of the spectrum from 0.4 to 0.7 micron. Notice that the visible energy from the 300 °K radiation curve is so low that it does not show up on the chart (300 °K is near room temperature).

Max Planck (1858-1947), a German physicist, studied Wien's technical paper and was able to improve on the accuracy of Wien's model of energy radiation (in a 1900 paper). Planck came to a conclusion which he himself found very hard to believe, which is that the energy changes come in tiny "bursts" rather than increasing or decreasing gradually. The quantum jumps of energy form the basis of *quantum theory*. He spent many years trying to show exactly what produced the corpuscular nature of radiation but was never successful. He was not able to apply "classical theory" (analysis) to derive the desired result. Classical methods use equations based on the rates of change of variables (primarily distance and force) and are used to form mechanical models, a system of analysis used for many years as it still is. Planck's theory was not only received coldly at the time, it was initially rejected by renowned scientists. His theory of radiation is now a cornerstone of modern physics. Planck received the Nobel prize for his efforts in improving upon Wien's model of radiation.

Following the efforts of Wien and Planck, the investigations of the radiation of energy broadened. Radiation of mechanical energy is based on an "ether" of gas, liquid, or solid to conduct mechanical waves. The mechanical analogy was extended to the radiation of light by the assumption that an *electromagnetic ether* was required for optical radiation to propagate. Two scientists, Michelson (1854-1931) and Morley, tested the ether theory at the end of the 19th century by testing for the existence of an "ether wind" created by the earth passing through the electromagnetic ether. An optical beam was reflected by a mirror such that two simultaneous beams were traveling along the same path but in different directions. A difference in travel times for the two beams was expected, but none was obtained. The experiment verified the constancy of the speed of light but did not solve the question of the existence of an ether. The failure of the experiment perplexed scientists, and other possibilities were considered. If the ether was dragged along with the earth, then these results would correlate with theory. However, the existence of the ether would also be necessary in the voids of outer space, and this possibility was not regarded as being likely.

Another milestone in science then took place. A new hypothesis was proposed by Fitzgerald and Lorentz (1853-1928) in which dimensions vary with the velocity of a moving object. The Lorentz equations describe the shrinkage factor as a function of velocity, and a later interpretation of these equations embodied the idea that the time interval measurements are different for the two beams. The Lorentz equations are an important factor in Einstein's special theory of relativity.

Albert Einstein (1879-1955) used the theories of Planck in his studies of the ejection of electrons from metal that was irradiated by light (photoelectric effect) and in 1905 proposed that light comes in "bursts" of optical energy, which he termed "photons," which are the jumps that occur in Planck's black body radiation model. Some of the leading physicists of the day initially thought Planck's idea to be crazy, but it was eventually accepted and Einstein received the Nobel prize for his efforts. The photon theory proved to be quite important, leading to many other scientific discoveries and providing a clearer conception of the characteristics of matter. He also later equated mass and energy (his now famous equation $e = mc^2$).

Then came another incident which threw the scientific community into disarray. Experimentation and measurement were becoming more accurate and more complex. New Zealand born Ernest Rutherford (1871-1937) was in charge of a laboratory in Manchester, England just before WWI, and in 1911 conducted an experiment, which consisted of directing a beam of alpha particles (subatomic particles of this type are present within the nucleus of atoms) at a thin metal film of gold, and found quite unexpected results. The degree of the effect of the foil on the beam was so very slight that it was totally contrary to the expected result, and Rutherford himself described it as the most incredible event in his life! Most of the particles passed directly through the foil with only slight deflection, while a few bounced back. The results did not correspond with earlier theories, and these experiments were not initially regarded as particularly important. If matter were totally solid we would expect most of the particles to bounce back or somehow cause some destruction of the sample. If it were partially solid, then a lesser effect would be expected. These results were confounding.

Rutherford proposed a new theory about mass and space that was so radical that his colleagues thought that he had gone mad. His measurements had led him to conclude that space is quite empty. Eventually his ideas were accepted and, along with those of others previously mentioned, are still in use today.

Rutherford's idea led to a new model of the atom based on the measured scattering angles. It is possible to picture the scattering angle by visualizing a rolling ball moving through a group of hills. If the hills are wide and close together the ball will not make it through but will be deflected, while very thin, tall hills will not impede the path of the ball unless it hits one squarely and scatters. The current model of matter is pictured as being composed of atoms and molecules. Molecules are composed of atoms, and these atoms consist of protons, electrons and neutrons. The protons and neutrons are at the center, or nucleus, of the atom with the electrons flying around them in accordance with the Rutherford model.

A new physical theory then came into being. The Danish physicist, Niels Bohr, established some of the basic principles of *quantum theory*, beginning in 1913 with his interpretation of the radiation from certain molecular substances. Discrete spectral radiation is quite different from black body radiation, consisting of specific wavelengths for a given material rather than the wide, continuous band of radiation from a black body. With quantum theory, the properties of atoms can be related to the spectrum of their radiated energy. Before Bohr's proposed theory, no one had been able to use classical theory to explain the spectrum of even the most simple atom, hydrogen. The reason is that no correlation could be found between the frequencies of radiation and any mechanical model (or any other model). He established a relationship between the radiated energy and the radius of the hydrogen atom that correlated with the spectrum of radiation of hydrogen. The radiation quanta of Planck and the energy concepts of Bohr form the foundation of quantum theory. While Bohr's methods produce successful results, they are abstract ideas as we will see in the next chapter.

As it has been said, "History is destined to repeat itself." An observation of the above events shows that one thing stands out clearly.

Each significant theory was not accepted at first, and the authors were often considered to be crazy. Galileo and Copernicus were denounced as heretics and fools. Maxwell's theory was not only not accepted in the beginning, it has never been proven. Nevertheless, we now know that it has been an extremely valuable tool in designing electrical and communications systems and is in wide use today.

New ideas do not come easy. Planck's ideas were rejected, and Rutherford was believed to have gone mad. Even Einstein's original efforts were initially rejected. However, these theories were finally accepted by the scientific community after they were found to correlate with the measurements. Planck showed that energy variations are not smooth, and certain prior theories had to be either modified, rejected, or ignored. Rutherford produced a new model of the atom with the proton in the center which has survived until today (with additions and modifications), but it also created new problems which are discussed in the next chapter.

Sometimes theories are abstract and may not completely make sense. They may also consist of individual concepts that in some cases do not agree with one another. Although such inconsistencies can be troublesome, abstract ideas and concepts often prove to be useful tools when measurements conform with the theory. A set of theories can form a *model*. Scientific models do not necessarily answer why something is happening, but simply describe the manner in which events occur. Planck's empirical model of radiation did not answer why radiation comes in tiny bursts, but it led to the concept of the photon following Einstein's efforts. Rutherford's model of the atom pictured space as being quite empty, but it provided no reason as to why the negatively charged electron does not fall into the nucleus of the atom which is positively charged and attracts the electron.

Today's accepted contemporary analysis method for atomic physics is quantum mechanics which has evolved into an assemblage of theories, many of which do not agree with one another. A major trait of quantum theory is that it is mainly abstract, while classical theory produces a mechanical model that is much easier to comprehend. Classical theory provides more information, but it, too, has its difficulties which will be discussed later. It is a common belief that classical theory cannot compete with quantum physics in describing the dynamics of matter. But quantum theory is not without its problems, and some of the complications will be reviewed in Chapter II.

Another thing that we can observe is that each of these theories have changed with time as other scientists have added to them. Aristotle was not completely mistaken in his reasoning for the systems that he considered, although his theory was quite incomplete. He probably did not consider what might happen by dropping two rocks and analyzing the results as did Galileo. Had he done so, he might have modified his theory accordingly. Neither was Ptolemy or Copernicus wrong, but simply in error; a slight error of omission. Few, if any, theories have not changed with time. The combined contributions of Maxwell and Einstein produced variations in Newton's theory which had been based on Galileo's experiments. As yet, Einstein's theories have not yet been shown to have any discrepancies, but on the basis of these considerations it will happen.

New information can change a theory by addition or modification. The theories of most of these scientists, including Aristotle, have been modified in one way or another. When new and more accurate measurements are taken, new factors must often be taken into consideration. Maxwell's equations can be arranged to show that electrical waves move at the speed of light. Lorentz analyzed Maxwell's equations and concluded that light travels at a certain specific speed. The result of his efforts produced an abstract model in which space and time vary with speed. Einstein theorized that <u>nothing</u> in the universe can travel any faster, and the consequences of this limitation meant that Newton's theory had to be modified such that mass, time, and space change as velocities approach the speed of light. All of these principles were embodied in his *Theory of Relativity* which relies heavily on the Lorentz transformation.

Looking back into the past, we can see that scientific methods began in the time of the ancient Greeks with their logical methods of analysis. No major scientific breakthroughs occurred for over a thousand years. There are undoubtedly a number of reasons why progress ground to a halt, including the impact of wars and politics. The next major advances took place after the alchemists learned about chemical characteristics (and kept them secret for most of the time). Secrecy impedes progress, and much of the contributions of the alchemists has no doubt been lost because of it.

Science depends, to a large degree, on the ability to make measurements, and the precision of sensing methods improved over the years. For instance, length had been measured by pacing off a distance foot-by-foot, and the length of the "foot" has now been standardized to a value that is more than a million times more accurate. As the accuracy of data improved, the theories improved, since variations had been clouded by errors in measurement, and today most of the measurements are made by highly accurate instruments. Instruments can also measure things that humans cannot sense, which extends our vision of physical events. The better the instrumentation, the better our view of the events that occur throughout the universe.

We now also have many better methods for analysis, developed over the years, now available to us. And with the advent of computers, the power of complex analysis has been extended greatly. We can now process huge amounts of data in a comparatively short period of time. Classical theory, an invaluable tool for the engineer, is seldom utilized in quantum physics, and both of these basic methods of analysis are highly valuable. With these changes, the evolution of science has greatly changed our vision of the universe. Our simple human view of our surroundings is no longer sufficient to describe the dynamics of physical events. With new computing tools and using new methods, our view of the universe will assuredly change.

I know of no way of judging the future but by the past.

---P. Henry

Nothing is so firmly believed as that we least know.

---Montaigne

CHAPTER II
Difficulties with Present Physical Theories

In order to solve a problem, it must first be defined. If a new scientific theory is to be proposed, then the problems with existing theories must be identified. Scientific text books often do not dwell on the problems with the physical theories they contain or the inconsistencies of scientific knowledge. Such an approach is not, however, incongruous with the state evolution of scientific methods. Scientists do not necessarily dwell on inconsistencies, as long as some degree of success can be obtained in applying theories to the analysis of physical problems. If the results are successful, then the methods will be adopted. Nevertheless, the lack of complete congruity between theories is a flaw that can hinder full scientific development. For scientists who must deal with the real world, problems must be fully defined in order to avoid misunderstandings and to provide useful solutions. In the search for the right answers, we will now attempt to propose the right questions.

New discoveries in science can yield new solutions to problems, in addition to extending the body of scientific knowledge. For the scientists at the beginning of the 20th century, these changes were particularly dramatic. They were dealing with situations in which details could not be viewed directly, and their tests and observations showed that the prior visions of both outer space and inner space (atoms) were both incorrect and incomplete. While the problems that they encountered were extremely difficult, the results that they obtained were nevertheless quite outstanding. Planck did not provide an answer as to why energy comes in tiny bursts, but originated an improved model of radiation. Einstein applied a method of reasoning similar to that of Planck, picturing optical radiation as consisting of granular bits of energy he called photons. Rutherford's model of the atom depicted space as being quite empty, and, although it provided no reason as to why the negatively charged electron does not fall into the nucleus of the atom which is positively charged, it has led to many other discoveries in atomic science. Bohr's abstract picture of the atom allowed the prediction of the spectrum of radiation of excited atoms. It is evident that each new theory is likely to be incomplete and have flaws.

In addition, sometimes theories do not completely make sense, and abstract ideas may have no direct correspondence with the real world. Bohr pictured the atom with the radius of the rotating electron reaching <u>any</u> size, and while we know that the orbital radius must be limited to a small value, his calculations have proven to correlate accurately with measurements of optical radiation of atoms. Theories may also consist of individual concepts which may be contradictory with one another. For example, mass consists of atoms which contain electrons, protons, neutrons and sub-atomic particles. Is the electron a particle or an electrical wave, such as a tiny radio wave? Physicists have found that it sometimes acts as a particle, sometimes as a wave. It would be very valuable to remove such inconsistencies which can cause confusion.

Also, ideas and methods may utilize concepts that do not exist in the real world. Mathematics, for instance, is an abstract science. Numbers do not exist except in man's mind. Mathematicians talk of *infinity* which is said to be any number divided by zero. The idea of infinity is highly abstract. Is one orange divided by no oranges an infinity of oranges? Can an orange can be divided into an infinity of oranges of zero size? Obviously infinity has its problems in the real world, and yet scientists put the concept of infinity to good use in utilizing mathematical methods to solve complex problems. Abstract analysis is used as a tool in complex analytical theory and is widely used in solving practical engineering problems.

It is necessary to use care in applying abstract ideas to the real world. Mathematicians are scrupulous in developing a basis of axioms and restrictions that allow the manipulation of numbers that does not result in ambiguous solutions. The engineer and physicist utilize these methods, but calculations involving "singularities" must be carefully applied. A singularity is a point in mathematical space where an infinite result at a certain point is possible. There is no firm evidence that infinity exists anywhere in the universe, and yet there are scientists who believe that singularities exist in outer space in the form of "black holes" or points of infinite energy and mass. Measurable evidence to support this idea is incomplete, and the theory is not fully supported by other physical phenomena.

There is considerable evidence to support the idea of a finite universe. Faraday showed that electrical charges vary in jumps, a process which was later attributed to the consequences of the electrical force of the individual electrons and protons which are all finite. Planck showed that energy is radiated in finite "quanta," thereby hosting the beginning of quantum theory. Einstein showed that light is radiated in finite optical quanta known as photons. Matter is now believed to be constructed from a finite number of atoms. Each galaxy in outer space contains only a finite number of atoms. All matter consists of a finite number of tiny building blocks. Can matter be condensed into zero space? A mathematical singularity would be necessary in order for such an event. But the requirement for zero space would necessitate the electron merging with the proton, and they would presumably annihilate each other. The only supporting factors for the existence of singularities are hypothetical. The evidence thus far collected indicates that ours is a *finite universe*, all theories to the converse notwithstanding.

We can no longer rely solely on our direct human senses, and we must now depend upon machines to do almost all of the sensing of physical phenomena. In the recent past, we have had much more information available to us than has existed in the history of man, due to the development of better machines and sensors, and there is now a great deal of recorded data to utilize for analysis. When it comes to atoms, it is difficult to visualize what is happening inside. While molecules have been imaged, pictures of atoms require much smaller optical resolution. No one has ever seen an image of an electron or proton because it is so tiny, and yet we produce machines that utilize their characteristics to create radio and television, the telephone systems, lasers, and many other functional devices.

The inventions of the microscope and the telescope greatly changed the vision of the universe. In order to obtain an image of an electron or proton, it would have to be probed by something smaller in

Ch. II - 34

size in order to obtain any degree of resolution. We cannot measure the volume of a thimble by trying to fill it with basketballs. For ordinary light, such as that radiating from our sun, many electrons would physically fit within the dimension of the shortest (single) wavelength of this type of radiation. Therefore, the length of these light waves is much too long to sense the construction within an atom, and the resolution from images illuminated by the comparatively long wavelength light from the sun would produce only a blurred image of an electron, even if a microscope with sufficient magnification to produce a visible image could be made. The electron microscope has much better resolution, but even it does not provide sufficient resolution, since the probing electron and the measured electron are of the same size.

The <u>time</u> that it takes to probe the electron is also a factor. The electron is moving extremely fast in its orbit within the atom, and the "shutter speed" of such an imaginary camera would have to be extraordinarily fast to eliminate blurs for an image of the orbit, perhaps faster than the speed of light (depending upon the speed of the electron). We will have to wait for extensive improvements in measurement techniques to be able to see what is actually happening within the atom. As these machines improve, it is inevitable that we will be able to sense phenomena never before observed, and so these future machine measurements will be changing theory in other ways, even if it is not possible to view the electron or proton.

The present accepted method of scientific analysis of matter is quantum theory. While Planck was the first to observe energy varying in tiny increments now called "quanta," Niels Bohr is credited with substantial contribution to the development of quantum theory because he was able to derive a method that predicted the wavelength patterns of energy emitted from excited atoms; something never before accomplished. As we reach the edge of knowledge, reality begins to get fuzzy since what is true is difficult to tell from the untrue. It is for this reason that truths evolve with time. Bohr believed that there is "no deep reality," and Bohr's theory that addresses atomic radiation reflects such a philosophy. His theory allows the radius of the atom to assume a dimension of <u>any length</u>, as long as the resulting calculation matches the measurements. He related the wavelength of radiation to the change in radius of the electron path within the atom. While his theory seems to work, we know that the wavelengths can be much too long to fit the actual radius of the atom, which does not permit a clear visualization of the true construction of matter.

Quantum theory is also based on the conservation of energy (and mass, since mass equates to energy). Restricting observations to only measurements of energy, space, and time is a highly reliable approach, due to the law of conservation of energy, but it also hinders visualization. Consider the difficulty in trying to construct the image of a car if we could not see it and could only measure the amount of energy it radiates and its speed.

On the other hand, the mathematics of classical analysis, which deals with rates-of-change of variables, is much clearer and permits a more detailed visualization of reality (an excellent discussion of reality is given in *Paradigms Lost* by John Casti). The problem is that classical analysis has not provided the proper answers for the process of atomic radiation, and, if these methods can be applied, the keys to the understanding of the universe may be found. A classical analytical solution method for atomic radiation will be presented in Chapter XIII. The results appear to be rational and comprehensible.

There are many other examples of how our visions have changed with time and discoveries. While the ancient Greeks conceived the idea of the atom, it was not until the developments in chemistry and physics that atoms were established as the building blocks of matter. Electricity, *the unseen power*, is a more recent development that has allowed transmission of power across the country through vast power networks, and, similarly, radio transmissions permit information (data) to be transmitted afar, even to distant planets. And yet we cannot see what is happening within the atom or view the transmission or radiation of power. In fact, most of what we now "know" cannot be seen directly. Therefore our newer vision of reality exists mostly in our imagination.

Another problem is that no one person can supply all of the answers to all of the questions of science. As a consequence, many diverse ideas have been proposed as to the makeup of our universe. There is also a trend, today, to sometimes propose sensational theories which have very little substantiating evidence. In cosmology, very little can be proven by measurement, and cosmological theories tend to be highly dramatic. The origin of the idea of a "black hole" in space is not clear. Some believe that it began with the mathematician, Laplace, in the late 18th century, who considered the gravitational attraction of a huge star dragging matter to its surface (*The Exploding Universe* by Nigel Henbest). The presence of a black hole in space, from which no radiation could escape, became more believable after Einstein proposed that radiation has a gravitational attraction to mass (scientists now believe that radiation can occur at the edges of a black hole). Cosmologists have been looking for black holes in outer space, and some evidence does exist which suggests their presence. Many of the recent pictures of the universe that have been collected by the Hubble telescope show grandiose displays of energy radiation, and it seems that most every new, unexplained phenomena is currently attributed to the black holes of outer space. Probably the strongest evidence is provided by measurements of bits of matter, moving at such high speeds around a location in space that only the huge gravitational force of a black hole could hold them in orbit. All of the evidence is not yet in on the subject of black holes.

New information can change a theory by addition or modification. Events of this type happened with most of these scientists from Aristotle to Einstein. New experiments are constantly being conducted, with greater accuracy of measurement, and new factors must be taken into consideration as new variations, not before seen, become apparent. We must be careful to avoid believing so strongly in any theory that we cannot accept that it may be flawed in some way.

Newton's ideas conform to most of what we encounter in our human experiences. These experiences are changing with the advances of science, as is especially true with the theories of Einstein. Newton based his theories on the observations of force, while mass and the photon were the fundamental quantities of Einstein's universe. His ideas must be taken into consideration when considering events occurring within the atom or in outer space, and for the transmission of optical data and radio waves, since optical beams travel at the speed of light. Einstein found it necessary to modify the spatial equations such that the dimensions of mass, time and space vary with the velocity of a moving object and used the works of others in establishing his theories. But what if mass is not the simple constant that it has been assumed to be? A more detailed description of mass could then change our vision of the universe immensely.

Einstein's *Theory of Relativity* is now an accepted theory, but there is no reason that it, too, cannot change. Is it really necessary to view time and space as being distorted by velocity in order to make observation correspond to theory? It would seem so, but other possibilities must also be considered. And what about the speed of light? Can nothing in the universe travel any faster? This possibility must also be allowed. If there is one thing lacking in Einstein's theory, it is what constitutes mass. Einstein went so far as equating mass and energy but was unable to determine how the forces that relate to mass are produced. He wanted to *"build a pure field physics"* but was unsuccessful and admitted it. He concluded that *"For the present we must still assume the existence of both: field and matter."* For him, the gravitational force was simply the consequence of a separate entity: *matter*. The difficult problem of the meaning of mass will now be unraveled, and this will lead to new thoughts about the makeup of the universe.

It is inevitable that the truths of today will not be the truths of tomorrow. This book is based on that principle.

There are more things in heaven and earth, Horatio,
than are dreamt of in your philosophy.

---Shakespeare

Science is not just a collection of laws, a catalogue of unrelated facts.
It is a creation of the human mind, with its freely invented ideas and concepts.
Physical theories try to form a picture of reality and to establish its connection
with the wide world of sense impressions.

---A. Einstein

She is spherical, like a globe.

---Shakespeare

CHAPTER III

Our Circular Universe and Rutherford's Atom

Many of the objects that we view appear to be circular, nearly circular, or parts of circles. The sun and moon appear to be round as do pictures of planets and stars. The horizon at sea appears circular. Our concepts of atoms and molecules are pictured as being circles or spheres, as are many subatomic particles, even though we cannot see them directly. The far off galaxies are viewed as spirals whose stars rotate in circles.

It is possible to picture the entire universe as being circular in nature. A circular universe is essential to our existence. A non-circular universe would be chaotic. Every day, week, year or month would vary in its period if the planets did not rotate and move regularly in circular paths. Clocks and calendars would be useless. Anything that is harmonic or periodic has a degree of predictability and dependability. If everything that we experienced was constantly changing, we would not be able to draw conclusions or provide proofs for any idea. Every experience would be a new one. Since human life is believed to exist due to an evolutionary process of development, it is not likely that humans, themselves, would have developed without harmonic motion.

Trees and rocks do not appear circular when viewed from a distance, but when their images are magnified the greater definition shows a much more complex structure. We now know that they are composed of atoms and molecules in various arrangements, and so they consist of assemblages of circular objects. Even a straight line drawn on a page is not what it first appears to be. A graphical line is far different from a mathematical line. The mathematical definition of a line does not exist, except in our imagination, since it has to have zero width, and we would not be able to see it. So all measurable lines have a finite width, and therefore straight lines appear as bars or rectangles. A line can be formed from a group of squares, and a square can be represented as a group of circles. Therefore all manmade objects are also circular.

To illustrate the above point, first fill a square with the largest circle that will fit within it. Then add another group of circles of the largest possible size to fill the voids. The procedure is continued until the circles become too small to see and the square is filled as is shown in Figure III-1.

Figure III-1. Decomposing a Square into a Series of Circles

The above mathematical procedure is one method of showing that any object can, in one way or another, be represented by groups of circles or spheres. While the method is efficient, it is not nature's choice.

Everyone should know and understand "sine waves." When a point on the circumference of a circle is rotated and measured along only one axis, the projection is a sine wave that undulates with time, passing from one extreme to the other periodically. A rotating circle has two projections on each of its two axes that are sine waves, and so a circle can be represented by two sine waves existing in two orthogonal (at right angles) planes as illustrated in Figure III-2. The sine wave on the right represents the vertical excursions of a point that moves along the circle, while the sine wave below shows the horizontal variations of the point. If the circle is rotating at a constant speed, then time is represented along the horizontal axes of the sine wave curves.

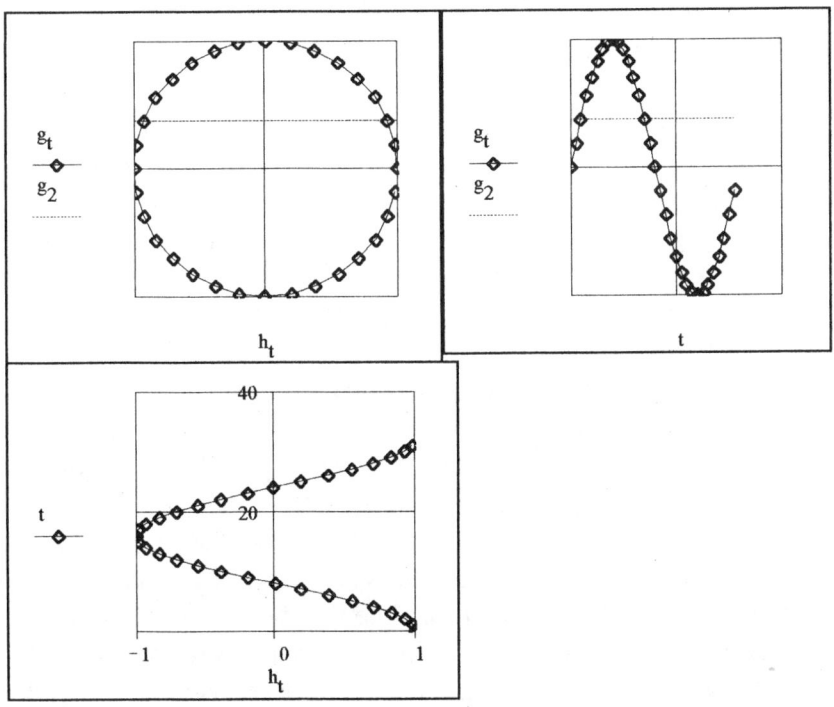

Figure III-2. Orthogonal Projections of a Sine Wave

A rotating square can be represented by a <u>time-series</u> of sine waves by a similar procedure. The decomposition of a function into a series of sine waves is called Fourier analysis after the French mathematician who derived the formulas for the series.

Any time function, such as speech, has a Fourier frequency spectrum consisting of sine waves of various audio frequencies similar to those shown in Figure III-2. Radio waves vary much faster than audio waves and therefore have higher frequencies. The AM radio band is centered at a frequency that oscillates at one million times per second, while our voice frequencies fall off rapidly above one thousand times per second. Frequency is measured in cycles per second (cps) or Hertz (Hz). Thus sound and radio transmissions both have circular characteristics, although the frequency spectrum of a radio wave is much higher than the frequency spectrum of an audio wave because the frequencies are higher. Anything that varies with time can be represented by a series of sine waves which are parts of circles and are therefore circular functions, and these characteristics provide useful information.

The entire universe can thus be represented as vast combinations of circular objects. Matter, however, is not formed by the basic geometry of Figure III-1 even though it, too, is circular. Nature has designed the geometry of the chemical elements (atoms) such that any piece of matter of any size can continually be subdivided into portions that have similar characteristics. In order to be able to continually subdivide matter, the smallest portion consists of one or more atoms arranged in a specific geometry. The atoms and molecules that make up solid materials arrange themselves into arrays that have voids between them. The characteristics of matter depend upon the way atoms are packed together, which will be explored in more detail in Chapter V.

A line drawn on a page would appear as thin rectangular bar whose surface contains atoms or molecules which transmit, absorb or reflect light. If we draw a very thin line that we can just barely see, it might be about 0.001 inch wide. But, if this line were only one atom thick and four atoms wide, it might appear, after extreme magnification, as in Figure III-3 (for atoms aligned as they are in a crystal).

Each atom of gold is about 2.5×10^{-8} centimeter wide, so if the line were made of gold it would be only 1.25×10^{-8} millimeter wide which is 80 million times thinner than a line that we can barely see. Our eyes do not have the resolution to see the fine detail of matter, and the lack of visualization of fine detail accounts for what we imagine to be straight lines which are really assemblies of circular matter. A similar situation exists for rocks and trees which are assemblages of atoms in the form of tiny circles or spheres.

Many materials are "amorphous," having structures that do not have orderly patterns. Water and wax are examples of amorphous

Ch. III - 43

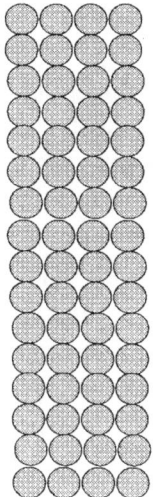

Figure III-3. A Line Composed of Atoms
(four atoms thick)

materials. Crystals are formed from atoms (or molecules) which are arranged in exact geometric patterns. Salt and sugar can exist as crystals. Ceramics are materials that consist of tiny crystals in a semi-crystalline form; partly amorphous and partly crystalline. The manner in which the atomic spheres are packed together determines the characteristics of the crystal. The tightest structure is called "closest packing," and each atom touches twelve of its neighbors. Silver, aluminum, copper, nickel, and lead have the closest packing structure. Not all crystals have closest packing, such as tungsten, chromium, and iron which only touch eight adjacent atoms. Molecules are groups of atoms, and so all material is composed of different three-dimensional geometric patterns of tiny spheres.

Events that we measure in time can also be considered to consist of circles or parts of circles. Sinusoidal functions that vary with time and space are called waves, such as the waves that can be caused by dropping a small rock into a pond. Optical radiation also consists of waves. The frequency of optical waves, radiated from material which is heated, can be measured, and the resulting spectrum consists of all of the frequencies being radiated. The interpretation of the resulting spectral data provides an important tool that is used in the formulation of atomic theory.

Everything that we see or otherwise sense consists of either matter or energy which comes from matter. The human body senses radiated energy by detectors that are built into our system. Optical energy enters our eyes and is sensed by tiny optical detectors, while audible energy is sensed within our ears by audio detectors. We also have the sense of "feel" which is accomplished through the force sensors in our fingers by pressing against matter. In order to comprehend the characteristics of matter, we must begin by obtaining information about matter and what we call its mass. Basically, matter consists of molecules which are formed from specific groups of atoms. The known atoms are listed in the Chemical Table of Elements. If we can first understand atoms, then we will be able to begin to understand matter.

So atoms are the building blocks of our universe. Within the atom are electrons, protons, and neutrons (and subatomic particles). How do we know that subatomic particles exist? We do not know it for certain, but bombarding matter with particles produces reactions that last for a very short period of time, leading to the belief that subatomic particles exist in the nucleus. The current model of the atom that is in use today is based on the basic model developed by Rutherford as described in Chapter I. The Rutherford model of the atom is surprisingly "empty." He concluded that, due to the narrow scattering angle, the diameter of the nucleus of a hydrogen atom is very small, only .01% of the diameter of the atom, and the electron is equally as small. A partially-scaled picture of the hydrogen atom, as defined by Rutherford's model is shown frozen in time in Figure III-4 and is in the form of a circle. If we had scaled the atom exactly, the electron and proton would be difficult to see. The volume of these particles takes up only a little more than a millionth of a millionth of the volume of the atom. Matter is nearly empty space! The distance between adjacent atoms and molecules is insignificant compared to the comparative void within the atom.

To picture a minuscule bit of matter more clearly, let us use an analogy of the thin gold foil with a unique small "forest" about 400 trees thick (one for each gold atom) with branches extending to a 25 foot

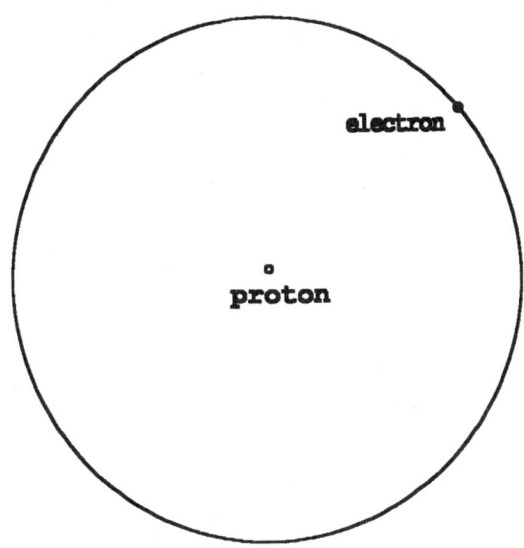

Figure III-4 Illustration of Rutherford's Atom

diameter. Such a forest would appear quite strange with only 79 leaves per tree which are 1/30 inch wide (to simulate the 79 electrons of the gold atom), and a tree trunk of nearly the same diameter (to simulate the nucleus of the atom). Obviously the forest would not be a very dense, more like thin bamboo reeds with very long, thin invisible branches with the trunks much stronger than steel. Actually, the leaves of our hypothetical trees would be rotating about the trunk at a high speed for a better analogy.

The above analogy is hardly perfect, but it is not likely that we would be able to find a perfect analogy of any human experience that is very close to that of the dynamics of atoms. The trees in our imaginary "forest" consist of electrical fields which have unusual characteristics that are difficult to visualize, and these fields are in motion, producing other effects yet to be discussed. It is an extremely harsh environment, having very strong forces far too strong and moving much too fast for us to easily picture.

The gold foil, which is only some 400 atoms thick, actually

passes some green light through it as is seen when held up to the light, but it is not highly transparent. Metals have less tendency to pass light than non-metals, and some non-metals such as silica pass light quite well. Since gold is a good conductor, it takes only a few atoms or molecules of thickness to impede light.. Silica is used in fiberoptic cables which can transmit light for as much as several miles without severe attenuation (reduction in energy), so the light waves must pass through several billion molecules before they exit the cable.

In general, gases are more transparent than solids or liquids. We can easily see through the lightest gas, hydrogen, and through our atmosphere which consists mostly of oxygen and nitrogen. There is considerable distance between these molecules in our normal environment (low pressure), and so there are few molecules to absorb energy, but not all of the wavelengths of light pass through any gas unimpeded. Light of certain wavelengths are absorbed more than others.

Rutherford discovered that the nucleus of the gold atom is not that much larger than the nucleus of the hydrogen atom. But the tiny particles within the atom are extremely high energy sources which deflect alpha particles (subatomic particles emitted from radioactive materials) when there is a collision. Due to the very high electric field strength within the atom, very strong forces and high energies exist, and they have internal electrical and magnetic fields which interact with matter and other fields passing through them.

Therefore matter, by Rutherford's model, is formed with circular or spherical building blocks which are mostly hollow but with very high pinpoints of high energy. The electrons fly around the nucleus at very high speeds with the protons and neutrons in the center. The speed of the rotating electron has been computed by utilizing Newton's second law and equating the centrifugal force with Coulomb's electrical force of attraction. The resulting electron velocity is slightly more than <u>two million meters per second</u> which is beginning to approach the speed of light at 300 million meters per second. Our human senses tell us that some of what we see is very solid and can be very strong such as steel, while instruments of much greater sensing capability show that all of space is mostly empty except for the strong rotating electrical fields within each atom! The concept of space as being empty does not present a contradiction but a phenomenon which deserves some discussion.

Human senses are extremely limited. We sense by feel, smell, taste, sight, and sound. Forces are sensed by the human body mostly by feeling and to a secondary degree by sight. Our nervous system is the electrical network that transmits the sensing signals to the brain. We now know that our human sensing system provides gross and inaccurate measurements using these few senses. Instruments, on the other hand, have a much greater range of parameters by which to measure events, such as optical radiation at many wavelengths, ion bombardment, electrical radiation at various frequencies, etc. Instruments are also usually much more sensitive than human senses. (The human eye has a range of detection of optical energy which extends over some seven decades (a decade is a 10:1 ratio) of range. Semiconductor diodes did not have as much range as the human eye until about twenty years ago, and they can now span over fourteen decades of range.)

Scientists have added further details and improved on the Rutherford model, but it has not changed radically. The pull of the rotating electron exerts some effect on the nucleus of the atom, and the resulting force is believed to produce a small rotating motion of the nucleus. So, for Rutherford's modified model, everything within the atom is moving in circular orbits of some type.

Most atoms are stable, in the sense that they do not emit energy, but there are <u>only some 109 different elements</u> plus various isotopes (an isotope has more neutrons in the nucleus of the atom). If everything in the universe consists of circles, or parts of circles, which can be analyzed mathematically, and if there are only a little over a hundred different elements, then what is so difficult about atomic physics? The task of analyzing matter is not easy since Mother Nature has done a good job of taking a simple situation and making it extremely complex. Suppose, for instance, that we have a tiny ball that would bounce forever once dropped and it is electrostatically charged so that it is attracted to another ball of the same type. Then take billions of these balls and start them bouncing and throw them together. The resulting conglomeration would tend to simulate, to some small degree, a one centimeter cube of matter. Now try to calculate the dynamics of the resulting mass. This calculation problem is a small effort in comparison to the difficulty of assessing all of the characteristics of matter. We will see that our most simple element, hydrogen, with but one electron and one proton, has very complex properties, and the various atoms combine in many ways to form molecules. And our earth has some 10^{27} (a billion, billion, billion) cubic centimeters of matter within it. The universe contains many stars, each of which is much bigger than the earth. Ours is a universe of large numbers.

As we learn more and more about the universe, new questions begin to arise. With all of the galaxies, stars, and planets in the great extent of the universe, why aren't there more basic elements (atoms)? What causes matter to radiate energy? Why do objects exhibit gravitational attraction? What is inertia? Even though our knowledge of physics has increased tremendously in recent years, these subjects have many mysteries associated with them since they are at the border of human knowledge. Some scientists have stretched their imagination to great lengths in probing the secrets of the mysteries of the universe. But imagination is a necessary starting point, since scientific development begins with supposition, which is then followed by substantiation. If confirmation does not follow, then other suppositions must be considered.

A general idea of the characteristics of matter and the universe has been pictured. A circular universe in every respect, and a universe that is based on the tiniest particles of matter. Hypothetically, if we could stick the little atoms together in the proper way, we would be able to construct a universe such as the one we have. The pertinent historical developments of science that have contributed to our present understanding of the universe have also been identified, so that the present state of our knowledge can be assessed by researching the efforts of others. Another major ingredient in our model of the universe is energy, and there is a lot of it out there. If we are to understand the universe, then we should have a feeling for the great power and <u>forces</u> that are acting around us. We will now examine an important contemporary theory which describes the events that presumably occurred in the formation of the universe.

*The Universe,
as far as we can observe it,
is a wonderful and immense engine.*

---Santayana

CHAPTER IV

The "Big-Bang" Theory

On a clear summer night, the brilliance of pinpoints of light from ancient stars (as much as 30,000 years ago) dot the heavens. The stars form patterns and move slowly in concert across the sky. Youngsters learn how to locate the Big Dipper, pouring its imaginary water into the Little Dipper, by recognizing its shape. Once the Little Dipper is found, it is then possible to locate the North Star, which is a more consistent method than a using simple compass to determine the direction of north (due to local magnetic fields that can be present). The sun is at the center of our solar system, and the light from other nearby planets, rotating about the sun, are produced by the reflection of sunlight from their surfaces.

We are in a galaxy, the Milky Way, which contains many stars. Most of the light from stars that we can see with our naked eyes comes from stars within our galaxy. The milky band that stretches across the sky comes from many far distant stars that are also within the Milky Way, too faint to appear as bright and sharp as the nearer stars. Our sun is not at the center of our galaxy, but far enough towards the center that we are on the inside looking out.

There are many other galaxies within the universe. We can neither see nor imagine the full extent of the tremendous action and power that transpires throughout the universe. Only a small fraction of one percent of the stars that have been detected can be seen directly. The telescope, however, reveals the details of distant stars and galaxies, and now we have fantastic pictures, transmitted to us from the Hubble space craft, that show phenomena never before witnessed.

While the Big Dipper is familiar to most people, there is another constellation that is even more easily recognized by the three stars in a row at its center. Astronomers list it as M42, and it contains a red supergiant star, a blue giant star, and a visible nebula (a large, glowing cloud of gas). The ancients probably spent more of their time viewing the stars in the sky, rotating slowly, but incessantly, around them. Their imaginations led them to visualize the constellations as outlines of human forms living in the heavens, and they named them after their heroes. M42 was called *Orion the Hunter* by the ancients and its center is sixteen hundred light years away.

After looking up into the night sky and viewing the wonderment, a question that comes to mind is "How did the universe begin?" That question has intrigued the world for ages, and many theories have been proposed. The "Big-Bang" theory, which can be traced back to the 1920's, finally became accepted by the scientific community by the 1960's since enough scientific investigation had been built up to support it. The Big Bang theory holds that the universe began with an enormous explosion, and that our universe is constantly expanding. The theory has been extended to include many sub-theories, some of the more recent ones (and even a few older ones) quite fanciful.

Scientists believe that the universe is expanding, which has been a major consideration in the formulation of the Big Bang theory. While measurements of the depths of the universe are, to say the least, difficult to make, strong evidence of an expanding universe nevertheless exists. The evidence, astronomical measurements collected by Hubble, is based on the Doppler frequency shifts of light emitted from stars. The direction that stars are moving to or from us can be measured using the Doppler effect. The Doppler shift of sound from a moving train produces a shifting tone for the whistle of the train, sounding as a higher note as the train approaches and lower as the train moves away. Light exhibits a similar shift in frequency, moving toward the red as the light source moves away and toward the blue as the source approaches.

Stars do have differences in color, sometimes appearing red and sometimes appearing blue. But the red or blue color can also be attributed to the color temperature of its surface, which is thousands of

Ch. IV - 51

degrees Celsius, as well as to the Doppler shift produced by star movement. This observational difficulty has been surmounted by studying the spectrum of the stars and looking for atomic "spectral signatures" (patterns of the spectral lines). Hubble was the first to measure the velocity of stars and galaxies by analyzing the spectrum of the optical radiation from stars and measuring the shift in wavelength of the radiation spectrum signature. The current scientific community now accepts the idea that the universe is expanding (sometimes called "inflation") because there is much evidence of this type that supports it.

 For instance, hydrogen is a substance common to stars, and its optical spectrum consists of a series of discrete wavelength lines (optical frequencies) which are unique to this gas. Measurements show that the spectrum is usually down-shifted towards the red, due to the Doppler effect, which would indicate that most of the stars *are moving away from us*. If it is true that most all stars are moving away from us, and if the universe is a shell, all starshine should show a shift towards the red ("red stars"), and the faster a star is moving away from us, the greater should be this red shift. The measured results almost always support the above suppositions. There are a few "blue stars," although, the blue shift of these few stars is very slight. If the star is moving laterally towards us, due to, say, galaxy rotation, then a shift towards the blue is allowable without disagreement with the bubble theory. It is also possible that a galactic explosion may produce a star motion that is in our direction, since it also produces a shift towards the blue. Since astronomers view what they believe to be explosions in distant galaxies, it is likely that such explosions add or subtract from the velocities of the stars, thus producing variations in Doppler shift.

 The cause of the huge explosion that began the Big Bang, assuming the theory holds true, may have been the collision of two bodies, but there is even one theory that holds that it was created from nothing. One of the pioneers of quantum theory, Pascual Jordan, has even suggested that a star might be created out of nothing! He reached this conclusion by equating the star's negative gravitational energy to its positive rest mass energy at the point zero. Alexander Vilenkin has claimed that creating something from nothing is a mathematical possibility, through the creation of an electron-positron pair that then annihilate each other. The sequence of events is said to start with one electron being created out of nothing, it travels forward in time for a while, then it turns around and travels backward in time until it finally meets up with its own creation (an excellent account of various theories of the creation of the universe can be found in *The March of the Big Bang* by John Gribbin). Is this a reasonable assertion or gobbledygook? George Gamow is said to have gotten Einstein's attention by his assertion that a star might be created out of nothing (during the WWII period). Concepts of this type are highly abstract and can be considered as more of an idea than a theory. The evidence of negative time has yet to be demonstrated. A single electron/positron pair does not have enough energy to create a star, and it would take many, many collisions for such an occurrence to take place. Scientists are obviously groping for ideas as to what <u>caused</u> the Big Bang.

Physicists and cosmologists have developed various models, based on quantum theory and the Theory of Relativity, to support the evolution of the universe as a consequence of the Big Bang. These theories are based more on *inner space* than *outer space*. The fundamental principles of atomic physics have contributed greatly to the development of such ideas. A popular concept is that in a very early stage of the formation of the universe, hydrogen, the basic element, was created. Only two basic particles, the electron and the proton, are necessary to create a hydrogen atom or a neutron. The neutron is by itself unstable and is not considered an element. It is attracted to a proton, creating a nucleon which is found in the nucleus of many elements (isotopes are elements with various numbers of nucleons). So all matter is basically composed of two types of electrical charge, the electron and the proton (neglecting subatomic particles, such as the hypothetical "quarks" within the proton). The construction of a neutron is less clear. More details on the construction of matter is given in the next chapter.

Out of nothing nothing can come,
and nothing can become nothing.
   ---**Persius**

Was hydrogen formed <u>prior</u> to the first instant of the Big Bang? Perhaps not. The following summarizes a somewhat different picture of what might have happened at the instant after the Big Bang:

> At the first instant of the Big Bang, a colossal display of energy resulted in the creation of tiny electrical charges flying outward. As the groups of electrons and protons expanded outward, additional collisions occurred, each collision producing more and more collisions. Due to the electrical attraction between unlike charges, they moved toward each other, eventually forming a tiny orbit wherein one moves about the other circularly. In this way, a basic particle, the neutron (which consists of an electron and a proton) was created in the form of a neutron gas. Neutrons tend to be unstable, except when they unite with protons within the nucleus of an atom, and some of the unstable neutrons decayed into protons and electrons. A proton and a neutron have attraction for each other and they formed a bond. The proton/neutron pair in turn attracted an electron, thus forming heavy hydrogen. From that point on, the other elements were formed by nuclear reactions (similar to current theory).

The difference between the above theory and current theory is the assertion that the Big Bang did not begin with a hydrogen reaction, but was the result of the meeting of two electrical <u>force fields</u>. The reason that such an alteration of current theory is proposed will become apparent in later chapters.

The method by which the universe is formed has been analyzed by various physicists and cosmologists, and various theories exist.

It went to pieces all at once,
All at once and nothing first.

---Holmes

George Gamow worked out the details of a nuclear process whereby a very dense neutron gas would create hydrogen and energy, which led him to the original idea of the Big Bang. His graduate assistant at George Washington University was Ralph Aher who worked out details of the model, the work being done during the period from 1940-48. Helium is created by the nuclear fusion of hydrogen, and so these two elements would therefore be the first to be created.

Most stars fuse hydrogen, of which they are mainly composed, into helium, as does our own sun. Helium will also burn, and older stars have more helium in them since the hydrogen is partly used up. The fact that 99% of the visible matter of the universe is made up of hydrogen and helium (both gases) lends credence to the Big Bang theory and the idea of a hydrogen furnace fueling it. Our expanding universe is constantly being fed by these gigantic hydrogen and helium nuclear fusion furnaces containing plasmas.

All atoms contain multiples of hydrogen atoms within them (in stable energy subshells), and astrophysical measurements indicate that most of the matter of space consists of hydrogen and helium. It is possible to construct any element from hydrogen by the process of fusion, wherein the protons and neutrons bond together within the nucleus and the electrons rotate about the nucleus. The reaction takes place as the temperature decreases, since the radiation that occurs at high temperatures knocks the electrons out of orbit. Therefore, it is believed that the heavier elements were fused at lower temperatures as portions of the universe cooled.

Various versions of the Big Bang theory have been proposed, and parts of many theories form a common basis which has been accepted by the scientific community. As a result, the mechanics of formation of the universe can be traced back to the first instant in time (10^{-43} second), which is such a tiny amount of time to be considered incredulous.

In the narrow span of time which we are currently spending our existence, the Hubble telescope is collecting what amounts to snapshot pictures of a portion of outer space where the events pictured occurred at an earlier time; a time nearer the time of the creation of our universe. The universe may, however, be much too large to be able to see very far along its expanse. The width of the universe depends on its rate of expansion, and to see the far reaches of the universe might require detection of light originating hundreds of billions of years ago (a very conservative estimate). For an expanding universe, the events that can be seen present a time-lapse picture of this expansion, and the data collected by cosmologists presents opportunities for analysis and contemplation:

> Our earth, along with the planets, stars, and galaxies, must have originated, in some form, at the location of the original explosion. The geometry of formation of our galaxy would have evolved during this interval of time, and we are viewing earlier geometries as we look out in space and back in time. If nothing travels faster than the speed of light, then the expansion of the universe must be occurring at no greater a speed. As all of these objects fly away from each other, they either generate or reflect energy in the form of light. This light is traveling at its maximum speed of 3×10^8 m/sec through space; a rate that is probably faster than the rate the objects themselves are flying apart.

If there was just simply one big explosion, it is possible that the matter of our universe is located on the surface of an expanding sphere (a shell) as pictured in the cross-sectional drawing of Figure IV-1. If our universe is in the form of a shell, then when we view distant galaxies, we are currently observing events that occurred at points in former time when the sphere had a smaller radius due to the time it takes for light to reach us. Therefore, the light which arrives from distant objects must have occurred when the sphere was much smaller in diameter.

If our universe is expanding at a rate approaching the speed of light, then this sphere will appear highly distorted and significantly smaller than the actual sphere, since the optical dimensions of the field of view at the opposite edge of the sphere would be shortened. If the rate of expansion is close to the speed of light, then the universe would appear more densely packed from our vantage point of earth (since the galaxies are further away than they appear), and the sphere as viewed <u>from our position on the surface</u> would appear distorted. The amount

Figure IV-1 A Cross-Section of the Bubble
of the Universe

of this distortion depends on how much the universe has expanded and the rate of expansion.

While there is evidence to suggest that our universe may be in the form of a shell, there is also other evidence to suggest that it may be of another form. Calculations, based on an estimate of the total mass of the universe, indicates that it is *flat*. All galaxies also appear flatter than they should be. So in what shape is the universe formed? If the universe is shaped like a shell, then what we see from our vantage point may be limited to a very tiny portion of an immense shell which appears to be flat. The limitations produced by our point of perspective make it difficult to reach a firm conclusion as to the exact shape of the universe, and it leads to other conjectures. While the earth appears flat on some locations on earth, we can easily see the curvature of the earth from an high-flying airplane or satellite. However, we have no vehicle of such comparative proportions to view the universe, and so the degree of flatness can only be substantiated by the degree of accuracy by which we can measure optical angles.

The theories of cosmology are probably the least proven of any science due to the proliferation of unknowns and the difficulty in testing the theory with a prediction followed by a measurement. Cosmological theories are developed in the form of models based on a hypothesis and then, if possible, synthesizing a small-scale reaction by means of an earthly experiment. If the theory cannot be tested, then there is room for doubt as to the validity of the agreement between the test and the hypothesis.

Not only do we have the limitation of having limited ability to conduct direct experiments to test theory, but time and space are believed to be warped in accordance with the Theory of Relativity. But is space really warped? Space must at least *appear* to be warped due to the fixed speed of light and our limited vantage point from which to make measurements. All such complicating factors must be considered in the analysis and are especially applicable for an expanding universe since all bodies are moving at high speed. If the universe is expanding at a rate near the speed of light, then current theories must be re-examined.

Supporting evidence for certain cosmological theories does, however, exist. Nuclear experiments have produced helium by the fusion of hydrogen, and this process is a key factor in some of the above theories about the formation of stars. Exploding stars have been witnessed, and spectral data supports the hydrogen furnace theory previously discussed (fusion of hydrogen into helium).

Belief in cosmological theories depends on what data is available and which appears to be the most sensible approach. Even the existence of other galaxies was not confirmed until the 1920's. We cannot look directly across the Milky Way since we are inside it, but astronomical measurements indicate that it, too, is flat but has a thickness. The measurements, at least for our galaxy, are consistent with the bubble theory, and the shell of the universe must be at least this thick.

The supposed flatness of the universe does not refute the bubble theory. From where we sit within the thickness of this (presumable) shell, we do not have a good vantage point to view the entire universe. If most of what we see in space is only our own galaxy, then how much of the universe can be viewed by the Hubble space craft? The amount of energy reaching us from distant stars depends upon how far away they are. It may be that the universe is so large that we will only be able to detect a small portion of the stars; those which are the closest.

The Theory of Relativity holds that nothing in the universe is faster than the speed of light, be it energy, field, or mass. This theory is based on observation and thus far no contradictions have been found by any measurements. However, let us tacitly assume that, notwithstanding this powerful theory, the speed of light can somehow be exceeded by some of the stars moving in space. Stuck on our little point on the huge shell of the universe, we would never see those stars which are going in the same direction that our galaxy is moving since the light is too slow to ever catch up with us. The same would be true for those stars moving away from us, so we wouldn't see any of them either. Such a result could account for the great amount of black space in the universe.

Evidence will now be presented that suggests the possibility that mass can travel at the speed of light. We will first accept the present cosmological theory of a flat universe. An apparent flatness would similarly occur when viewed from the location of a huge shell if it were not possible to see all portions of the shell. The flatness of the universe is therefore supportive of the bubble theory. It would also appear to be less flat than it actually is due to the large dimensions and the distortions caused by its rate of expansion versus the speed of light (the universe would be smaller at an earlier period in its expansion). A rough calculation of the present size of the universe, based on the shell theory, will now be performed.

A partial view of a shell is pictured in Figure IV-2. This shell represents the extent to which we can view the universe with our most powerful telescope. The amount of curvature of the shell is exaggerated for illustration purposes only. Presumably, we can only see a much smaller portion of the sphere. As the arc length becomes shorter and the shell thickness is reduced, the section of the sphere will appear to be a flat, similar to a disc. If all of the stars that we can view were to be plotted on a map, and if they fall within a shape such as this, then our

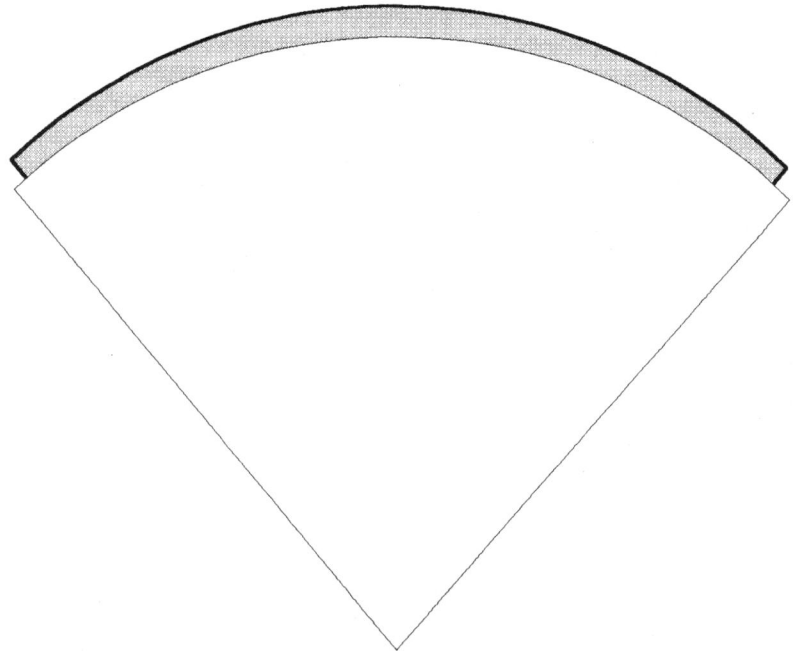

Figure IV-2. A Portion of the Surface of the Universe

universe may indeed be in the shape of a shell. If we had a map of the universe we could then reach firmer conclusions as to its exact shape. In the meantime, we will assume that the universe is in the shape of a shell. The directions of expansion are shown in Figure IV-3. The two furthermost points on the sphere are selected for the measurement. The angle between the radius and this chord is termed ϕ, while the angle subtended is called θ.

The Hubble Doppler measurements show that remote stars are moving away from us a rate which is proportional to the distance from the earth. The radial (outward) velocity can be much higher than the the tangential velocity along the bubble of Figure IV-3. Based on the assumption that these speeds add vectorially, the size of the universe

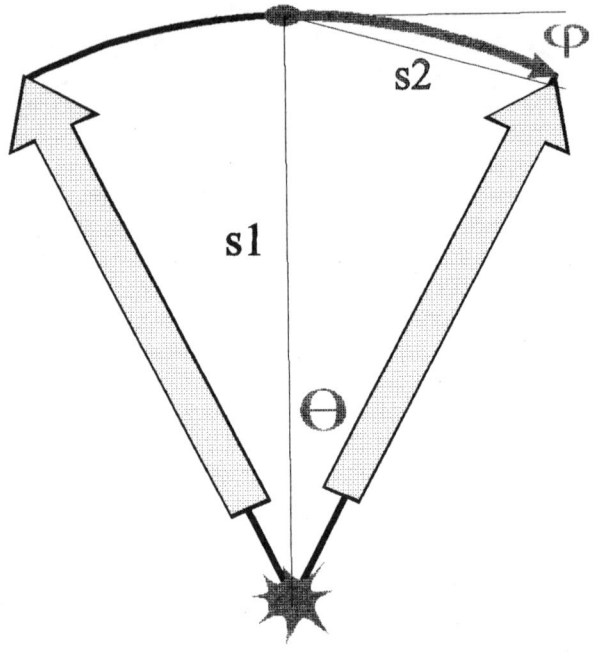

Figure IV-3 Expansion Velocities of the Universe

and the radial speed of expansion of the shell can thus be calculated:

$$v_t/v_r = \sin \theta$$

where

$$\theta = (s_2 \times \phi)/s_1.$$

The distances, s_1 and s_2, in the last equation are also assumed to be proportional to velocity, and <u>s_2 and v_t are selected from the points on the Hubble curve</u>. We do not know how flat the universe is, so we will estimate a curvature of .01 degree. These calculations portray a universe that is expanding at a rate which is almost at the speed of light, and the diameter of the universe is about 4×10^{27} meters. If the curvature is less than this amount, then the speed of light is exceeded. The distortion, produced by the speed of light affecting the measurements, increases the calculated radial velocity even further since the curvature of the universe appears greater at higher velocities.

The age of the universe can be calculated from these figures. For an outward expansion velocity of the universe of 2.4×10^8 m/s (0.8 times the speed of light), the age of the universe is 8.33×10^{18} seconds. There are 3.145×10^7 seconds in a one-year period, so the age of the universe is therefore 3.3×10^{11} years. The age of the universe as calculated by other scientists is in the range of 10 to 20 billion years which is about one tenth of the age that was calculated using this very simple method. (Note that the velocity of expansion was assumed to be constant after the Big Bang).

Is the above estimate reasonable? If the present estimate of the age of the universe is correct, then our universe is more curved than would be expected from the above calculation of the flatness of the universe. On the other hand, if the calculations are correct, then the universe is bigger and older than the present estimate. The universe is currently being mapped using more accurate techniques, and we will probably know more about this within the next several years. In any case, it appears that the universe is flying apart at speeds which approach, to some degree, the speed of light.

The above model of the expansion of the universe could conflict with the Theory of Relativity. If the universe is flatter than for the above estimate, then the universe is expanding at a velocity that can exceed the speed of light, which is another dilemma. Either the speed of light can be exceeded or it cannot. It is not possible to prove that something is impossible; only the contradictions to a theory can be elucidated. If two possibilities contradict one another, then the most likely possibility can be chosen. If we hold to the proposition that it *might just be possible to exceed the speed of light* in some way, then we need to examine any contradictions and eliminate them. We will need to take a closer look at the Theory of Relativity if we are to pursue this issue, and this will indeed be the case.

Another observation about the velocity of expansion is in order. As was stated earlier, much of the universe appears to be black. If our shell of a universe contains stars throughout its thin section, then we should be able to sense the emission of light in all directions, even if the stars lack definition due to their extreme distance. In other words, we should be getting light from the opposite inside edges of the shell if it is not moving away faster than the speed of light. However, if opposing portions of the shell are moving away from us at or above the speed of light, then the light rays would never reach us. The fact that much of the light is traveling too slow to reach us could account for the blackness of the universe that we see in the sky. The present belief, however, is that the inside of the shell is filled with unseen matter that absorbs light. A test for the absorption of the dark matter of the universe would resolve this dilemma. How far away is this dark matter? Doesn't it permeate the universe so that any spacecraft can grab a handful? Or is it always beyond our reach so that nothing can be proven?

It is now a well-accepted belief that matter was formed from hydrogen, and the elements of the universe evolved in the reaction process. In the larger stars, energy from the hydrogen/helium furnace is used to form the heavier and more complex atoms. The forces of such reactions are extremely high, and huge energy exchanges take place. There is another force which, although almost insignificant by comparison, plays a very important part. It is *gravity*.

Gravity has been an essential ingredient in forming the universe, even though it is a comparatively weak force. Larger objects are attracted to each other with much greater force, however, due to the aggregate attraction of many atoms. Thus the atoms collect in particles, particles form bodies, and in some cases they eventually form large stars. The larger bodies also have attraction for one another, the stars forming galaxies. Current theory takes us to this point without much controversy, but another question then arises.

Why did the stars and galaxies come together forming one large congregation of moving bodies? The expanding shell theory provides a reason here, too. The forces of expansion may be too high to permit aggregation if the gravitational force cannot overcome the force of attraction. All objects are flying away from one another, and the gravitational force decreases with distance, so the larger and closer objects are those most likely to come together. To test this assertion, the Doppler shifts of two large stars that are close together could be compared to those of stars, of the same size, spaced further apart. The Doppler shift can then be correlated to the velocities in order to see if a the stars that are <u>closer together</u> are not expanding outward as fast as the others.

A set of models has been developed that presents the means by which the atoms and molecules of the planets and stars were created. These models describe the formation of the universe back to the beginning of time.

So just how was hydrogen formed? The basic material of the universe, hydrogen, could have been formed before the beginning of the universe or afterwards. Unfortunately, this dilemma is not easily tested. Let us assume that it was formed afterwards in accordance with present belief. In the very early stages of the universe, it is believed that matter was in the form of a plasma, consisting of electrical particles. A possible mechanism for the formation of hydrogen is an electrical reaction or storm. After the initial collision of the Big Bang (cosmologists talk about the first 10^{-43} second of the reaction), the laws of physics must have governed the attraction and repulsion of all of the electrons and protons. Unlike charges attract, and the electrons and protons moved towards each other, forming the dynamics of stable orbits, unstable spirals, etc. by their mass, energy, and spatial positions.

The characteristics of radiation from *larger stars* indicate that they contain atoms which have grown to a large size. Large atoms collect numbers of pairs of electrons and protons plus some neutrons, reaching a size so large that these atoms have become unstable and begin to break apart, causing further energy generation. Instability of this type is called *radioactivity*. These energy-emitting particles merge and after some evolution form into spheres which appear as a tiny star.

As the radioactive atoms lose some of their electrons, protons, and neutrons, other elements are formed from the residue (uranium, it is believed, eventually turns into lead). Eventually the outer layer of material becomes immutable, with atoms and molecules locked together, and becomes a planet. This presents a picture of our own earth, which is believed to have a radioactive core and a liquid layer beneath a surface crust that floats upon it.

Are such theories believable? It depends upon what one will accept. Some require more proof than others. If we had all of the necessary data, it would be possible to plot the true origin and development of the universe. Unfortunately this is not yet possible, and theoretical scientists must use their imagination and intuition, and, using the tools of their background and experience, devise theories and then test them for successful results. Some scientists, such as the physicist, Bohr, and the English mathematician, Stephen Hawkings, have relied on abstract mathematical methods. In this book, the attempt is made to avoid using abstract ideas, except as a convenient analytical calculation procedure. Therefore the idea of infinity, a mathematical abstraction, is not fundamental to any of the arguments which follow.

The strength of the Big Bang theory is that much of what can now be sensed about our universe fits the theory. From all of these models, we can visualize the destiny by which our universe came to be. Only tiny bits of particles and/or energy (in large numbers) were needed to begin the process. The minuscule electrons, protons, and neutrons have very great attraction for each other and are fated to spin in space as various forms of matter. If an atom gets too big, then it loses some particles, and energy is released. The result is a coherent yet chaotic universe with huge exchanges of energy in various locations.

As with most cosmic theories, however, there can be unknowns, contradictions, and inaccuracies. Several of the gray areas of the Big Bang theory have already been mentioned, such as the flatness of our universe and the large percentage of darkness. A slightly different approach or estimate can produce models which differ greatly from those currently accepted. Is the universe expanding at or above the speed of light? It is doubtful than many would believe this to be the case, simply because of their belief in the Theory of Relativity and the idea that nothing can travel faster than the speed of light. But absolute velocities are not covered by the Theory of Relativity; only relative velocities. Our location within the universe does not permit us to accurately determine the speed of expansion of the universe since we are speeding outward, along with everything else around us. A map of the universe, created on the basis of data collected by the Hubble space craft may provide sufficient accuracy to be able to more accurately determine whether or not the bubble theory holds, and if it does, then the speed of expansion.

For the counter-theory, let us assume a fixed position in outer space and view the spectacle as an outside observer. Applying the Theory of Relativity, the dimensions of the moving body decrease as the speed of light is approached. Round objects traveling in a transverse direction would appear squashed. The diameter of the sun would be zero if its total speed vector were at the speed of light, and it would become invisible. Other objects which are traveling slower would have a finite width. Would our sun still radiate energy, and if it did would this energy ever reach us at our hypothetical vantage point?

Can these enigmas be eliminated? Consider the possibility that there is no limitation on the speed of objects, and that the apparent limitation is really a *measurement problem*. The effects of the speed of measurement can have a significant effect on the accuracy of a test. What you see may not be what you see. The errors produced by an insufficient speed of measurement is called the "sampling" problem and is well known. The possibility of a measurement problem will be discussed in more detail in a later chapter.

There is yet another known major problem with a cosmological theory, this time with galaxies. Einstein equated inertial mass to gravitational mass. Surprisingly, the planets within each galaxy are all moving at about the same speed. According to Newton's theory of gravitation, however, the planets nearer the center of the galaxy should be rotating faster than the planets at the edge of the galaxy. This dilemma has perplexed cosmologists, and some have even questioned the accuracy of the theory of gravity because of it. We shall return to this question further on in the book, because there is other evidence to indicate that the inertial force is not exactly the same as the gravitational force and that other problems with gravity exist.

The Big Bang Theory will not be investigated to any greater extent. To go further requires many more assumptions, and we have already used more of these than is desirable in order to make the theory believable. It is possible that the future mapping of the universe will prove that the bubble theory is incorrect, and that a new or different picture of the universe is required. If, on the other hand, the map pictures a bubble, then we will be able to determine the size and age of the universe and see how it compares with the above estimate.

Our look at outer space shows that it is our perspective within the universe and the accuracy of measurement that limits the possible assessment of the construction of a proper model of it. Our brief investigation illustrates the futility in looking outwards to find the answers we seek. The current state of the Big Bang theory holds that the universe consisted of plasma at a very early stage of development, and that atoms were formed from this plasma. It follows that we should study how atoms are constructed in order to understand matter. We will therefore reverse our direction and investigate the inner microcosm of atoms. We cannot see within atoms, even with the most powerful microscopes, and we must rely on other types of measurements and ingenuity. On the other hand, the experiments are controlled, and there is much more available information to draw upon.

The bottom is out of the universe!

---Kipling

For the proverb saith that many small maken a great.

--- Chaucer

CHAPTER V

Inner Space

In earlier chapters, a summary of the effects of gravity and the history of the scientific development of the theories on the subject was presented. We have seen that modern physicists and cosmologists have utilized atomic theory and the Theory of Relativity in developing their concepts of the universe. In the previous chapter, we found that the reaches of outer space, and our location in it, do not allow us to make the necessary tests to properly exercise theory. We will have to utilize the earth as our laboratory and examine the constituents of matter, and we will begin by examining the most basic elements.

The first step is to gather information and to do it with an open mind. We will use this information to form conclusions and eventually make some new assertions. In order to understand our universe, we must observe carefully what is happening in inner space. The overall scientific knowledge about atoms and molecules can be utilized to provide information for reasoning the causes and effects of events.

We will probe into atoms to see how they are constructed. It was previously proposed that everything in nature can be represented as being circular or composed of circular or spherical geometries. Rutherford's model of the atom was portrayed as being quite empty but having a high-force center which is produced by protons and neutrons in the nucleus.

The tiny electron and proton are attracted to one another by an electrical force, while the neutron is electrically neutral but is attracted to the proton. The electron orbits around the nucleus, attracted by the proton, but it is repelled when it gets too close to the nucleus. Laboratory tests and observations indicate that the electron is not bound to the nucleus nearly as well as the nucleus is held together. The protons and neutrons have much greater binding energy to each other, and much higher energy particles are required to break the nucleus apart than is necessary to remove an electron.

The structure of the <u>hydrogen</u> atom that was presented earlier was based on the rather straightforward Rutherford model. Many of the recent physical theories that have been proposed are based on sub-atomic particles and high-energy wave functions. Sub-atomic particles will not, however, be considered in the analysis. It is very difficult to probe the nucleus of an atom since a much smaller particle is required to produce much accuracy. Even less information is available on the internal construction of the nucleus, even with all of the data that has been obtained on subatomic particles that are ejected when it is bombarded. The unnecessary unknowns that exist far within the atom will be avoided, and the solid data at hand will be examined. The analyses will therefore be based primarily on the electrical charges that exist within atoms and their effect on physical reactions.

There are over 100 more different elements in existence, each of which contain various numbers of electrons, protons, and neutrons. All of the protons and neutrons of each atom are located in the nucleus, with the electrons revolving in various orbits around them. It is a curiosity that there only 100+ elements when there are so many electrons and protons available for the formation of matter. The characteristics of atoms have been investigated by countless numbers of scientists from various disciplines. We now know that the atom becomes unstable and tends to come apart when the collection of electrical particles becomes too high. Therefore, the heaviest elements are unstable (radioactive), and the atoms of stable elements contain a limited number of elementary particles. The matter of which the universe is formed seems to be of simple construction since there are so few basic elements upon which it is built. It isn't.

We do not yet fully understand the element hydrogen, even though it has but one electron and one proton. Neither electrons nor protons are particles as they are often regarded. Both have electrical fields that extend out into space indefinitely. As the electron rotates about the proton, it creates a rotational wave (a moving field). Although we have a set of equations which describe the properties of electrical and magnetic fields (Maxwell's equations), we still do not fully understand everything there is to know about electrical waves.

No satisfactory explanation has been offered as to why the electron is not drawn into the nucleus of the atom or why it is repelled when it approaches too close. Radiation and absorption of energy occurs in tiny jumps, as was shown by Planck, and each jump in radiation produces a photon, the tiny bit of optical energy (light) that travels at the maximum speed of anything in the universe (according to Einstein). The radiation spectrum of hydrogen does not correlate with other types of electromagnetic radiation that man has been able to generate. These enigmas are quite confounding since we cannot relate the elementary processes of matter to human experiences. Einstein, you may recall, pictured himself riding through space on a photon in developing his ideas for the Theory of Relativity, and he was quite successful in using this approach.

Bohr came up with a method of explaining how radiation is related to the orbit of the electron of the atom, but his thesis permitted an arbitrarily large orbit which is not rational. It is possible to view the hydrogen atom as a dynamic, nonlinear system and then employ the engineering techniques of classical analysis. Nonlinear classical analysis, unfortunately, is usually quite complex, difficult, and time-consuming; so complex that solutions often cannot be found. It is quite remarkable that these two little electrical particles (we need a better word for them) could cause so many problems.

All is not lost, however. Many contributions to the understanding of atoms and molecules have been made within the various scientific disciplines such as chemistry, physics, and engineering. The results of past studies and analyses has resulted in a model of inner space that allows predictions of results in chemical reaction, mechanics of physics, nuclear reactions, and electrical and electronic fundamentals which all fit together rather nicely. Today we see the fruits of these efforts in the fields of plastics technology, communications, television, manufacturing, energy, transportation, etc.

Good individual theoretical models of atoms have been developed in physics, but we do not yet have a good all-encompassing theory, but rather a group of theories, and sometimes these theories contradict one another. We need to look at portions of these models in greater detail in order to form some important conclusions. As the elements increase in size and complexity, the results become more and more difficult to describe. We will begin with the hydrogen atom which has been a common starting point for scientists since the constituents of hydrogen are common to all of the elements.

Hydrogen is an optically transparent gas. The hydrogen atom, the smallest and lightest of all of the elements with only one electron and one proton (except for *heavy hydrogen*, Deuterium, which has an extra neutron in this isotope), rises in air due to its low specific gravity. Two hydrogen atoms group together to form a molecule, and this molecule has great agility. Its average speed at room temperature and sea level is about Mach III, faster than a jet plane. The electron rotates about the nucleus in a tiny orbit, and a positive electrical force in the nucleus holds it in its orbit.

Application of an external electrical force across the atom can pull the electron out of its orbit and create an "ion," the negatively charged electron, and the positively charged proton (the isotopes may also have one or two neutrons in the nucleus). It has been determined by experiment that it takes 13.6 Volts of electrical force to pull one of these electrons out from a single atom (requiring an energy of 13.6 electron-Volts, (eV), to do so). Whenever the electron is removed, energy in the form of light is emitted. Light is produced over a range of frequencies (a spectrum), and a higher frequency (lower wavelength) corresponds to a higher energy or a greater eV value. It was also discovered that light was emitted when only 10.2 eV was applied to the hydrogen atom, but the electron was not freed from its rotation about the nucleus. More light was generated as the voltage was increased, but emission occurred in decreasing incremental steps as the 13.6-Volt level was approached. Thus light is emitted at various energy levels which relate to the position of the orbiting electron in the hydrogen atom, and each energy level corresponds to a spectral line.

In 1888, a scientist named Balmer studied the visible spectral lines of hydrogen and derived an empirical formula for the wavelengths produced. He found that they occurred in discrete intervals. Planck, in 1900, formulated the theory that energy in radiated or absorbed in specific energy units (which were called "photons" following Einstein's later efforts). In 1913 a Danish physicist, Niels Bohr, merged the two theories. Bohr reasoned that the electrons jumped to new orbits, one for each photon, and that the orbits must be integer functions (divisible by an integer). All energy jumps ("quanta"), in all orbits, in all elements, take exactly one or more photons. Bohr's formula, however, was not quite accurate for all elements, and there were small discrepancies in the predicted orbits. These discrepancies were resolved by a slight correction in the formula. Bohr's equation provided the relationship between the energy exchange and its spectrum, and he is credited with the establishment of *quantum theory*.

The correction factor that was used for Bohr's formula involved replacing the mass of the electron with a slightly lower mass termed the *reduced mass*. The reasoning was that, although the mass of the nucleus is much higher than the mass of the electron, the electron pulls the nucleus with it to some degree. Therefore, the nucleus would also revolve in its own tiny orbit in synchronization with the electron sort of like an unbalanced dumbbell, resulting in a slightly smaller orbit. So the mass of the electron was replaced by a slightly smaller "reduced mass" in the Rydberg formula which correctly relates energies and wavelengths.

The modification of Bohr's model illustrates some of the difficulties that occur in the concept of "mass." Mass varies with the movements of the orbiting electron and proton. The Bohr model lacks reality since it results in electron orbits which can have any dimension, and we know that is not possible since atoms have very small dimensions. Although it is an abstract mathematical model which does not fit all of the facts, it has proven to be quite useful in analyzing radiation since it produces accurate correlations with measurements.

Photons can be absorbed by atoms and molecules, resulting in either new electron orbits or ion generation. Conversely, photons can be generated when electrons move to a different orbital radius. Examples of devices that exhibit photon exchange are the LED (light-emitting diode) which has become commonly used for displays and indicators in electronic equipment, and the photodetector diode which produces electron flow when radiated by light. Ions can also produce light when they strike an atom or molecule. Light of a discrete wavelength can also be generated in a crystal which is irradiated by light of another wavelength.

Energy is transferred back and forth between matter in the form of dynamic collisions of atoms, molecules or ions, or by excitation by optical radiation (light). Reactions that seem smooth are actually grainy (to a fine degree), and this graininess is of a circular nature due to the orbital variations of electrons in atoms. Another complication is that probabilities enter into such reactions (Heisenberg's *uncertainty principle*). The momentum of moving electrons do not produce exactly the same result each time that they interact with matter and the parameters of the atom have a degree of uncertainty about them.

The above descriptions provide some degree of understanding as to the basic nature of atomic orbits and optical reactions, but what is the exact characteristic behavior of the *forces* of the very basic, simple hydrogen atom in space? Quantum theory is based primarily on energy rather than force, and the details of the action of dynamic <u>forces</u> is more difficult to describe by this type of analysis. If we can fathom the hydrogen atom, we can begin to comprehend the characteristics of other atoms, which means that we will have insight as to molecules, collections of molecules, and therefore matter itself. It is for this reason that the hydrogen atom has been highly studied by physicists, and why rich rewards in knowledge have resulted.

In order to understand how matter behaves, we need to understand the forces acting on the atom. Atoms and molecules behave differently when they collect in masses as opposed to the behavior of a single atom. The proton of the hydrogen atom is in the nucleus, and it holds the electron in its orbit, rotating in a very tiny circle. Opposite charges attract one another, while like charges repel. Therefore, the electrons of adjacent atoms will tend to avoid one another due to the repulsions of the outer electrons. If the electrons come too close to one another, then energy exchange will occur.

Let us consider the analogy of the hydrogen atom to a rotating tire on a car. If we were to take an unbalanced tire and cause it to rotate in space, the tire would wobble around in a circle about a center offset slightly from the hub of the tire. With the electron rotating about the proton, the hydrogen atom reacts in a similar manner. The individual hydrogen atom can therefore be described as a <u>mechanically</u>

unstable atom due to its non-symmetry and rotational vibration (this terminology is not to be confused with that of nuclear instability where the nucleus begins to disintegrate along with its associated electrons, producing radioactivity). When other similar atoms are nearby and two atoms collide, a portion of the energy of one is imparted to the other and vice-versa (energy exchange of this type takes place for all atoms, regardless of whether or not they are stable).

The atom can also be considered as an electrical oscillating system, the orbital rotation lasting for many human lifetimes (without an energy exchange between atoms). A common phenomena of electrical oscillating systems is that they tend to lock together when are at the same or nearly the same frequency and their forces are coupled together. In this case, the electrons and protons of one atom have attraction for the protons and electrons of the other atom, and two atoms will vibrate in synchronism with one another, thus forming a stable molecule consisting of exactly two hydrogen atoms.

A "mechanically-stable molecule," as defined here, is one which is well-balanced in its rotation and does not tend to unite with other molecules, and molecules of this type are therefore less likely to form solids or liquids. We can visualize the two electrons (of the two-atom hydrogen molecule, H_2) rotating opposite each other around the nucleus of two protons in the form of a balanced dumbbell. Protons, however, tend to repel each other, and so they will be slightly separated, in which case they will also be in motion, to a slight degree, about the geometric center of the molecule. Unbalanced molecules are mechanically unstable, tending to unite with other similar or dissimilar atoms or molecules to form stable molecules. The comparatively stable H_2 molecule has less tendency to unite with other atoms or molecules than a single hydrogen atom. But if it can be torn apart by the proper form of energy exchange, and if other types of atoms are nearby, then a violent reaction can occur. A reaction of this type occurs when hydrogen and oxygen are present and the gases are ionized. A water molecule is formed as energy is released, and the reaction can become violent (a hydrogen torch works in this manner).

To visualize the action of a single hydrogen atom in open space, picture a helicopter with an unbalanced rotor. If enough of one side of the rotating blade is clipped off, it will vibrate badly and eventually violently unite with the ground. If the propeller is counterbalanced with another rotating mass, it will then begin to run smoothly. Similarly, once the molecule is balanced it begins to react as would a billiard ball, bouncing off other balls but not uniting with them. Therefore, as these molecules bound through space and collide with other molecules, they have little attraction for one another and simply rebound at high speed. That is why the light hydrogen molecule can reach a speed of one mile per second or 3600 mph at room temperature and sea level. In this (standard) atmosphere there are about 11 billion collisions per second for each molecule.

The hydrogen molecules are so small and of such velocity that they can permeate rapidly through porous materials, such as certain ceramics, and they are lighter than air and tend to rise (the German zeppelins of the 1930's were filled with hydrogen). The average distance between molecules is more than 20 times the diameter of each molecule, which is one reason why hydrogen gas is very "open" and visibly transparent.

We now have some idea of how our lightest element functions in inner space so let's proceed to the next element, helium. Helium has an additional electron with two protons and two neutrons in its nucleus. The proton and neutron also are attracted to each other, and this pair, referred to as a nucleon, are commonly present in the nucleus of heavier atoms. With the two electrons revolving symmetrically about the nucleus, the helium atom is quite stable. Two helium atoms unite to form a molecule, which makes the molecule symmetrical in all three physical dimensions, thus causing it to be even more stable. It has very little tendency to unite with neighboring molecules due to the rotational stability. Helium is the first (and lightest) of the noble gases in the Periodic Table of chemistry.

The molecules of the noble gases are all extremely stable and have very little tendency to unite with other atoms or molecules and are therefore considered inert. It is for this reason that airships were filled with helium rather than hydrogen (the explosion of the German dirigible, the Hindenburg provided a graphic example of the dangers of hydrogen). Thus the mechanical stability of an atom is directly related to chemical stability, and a stable atom has little tendency to unite with other atoms.

Ch. V - 75

When we compare the helium molecule with the hydrogen molecule, it might seem surprising that two of the hydrogen molecules do not unite to form H_4 which would have high rotational stability and be chemically inert. The nucleus of the helium atom must, however, contain one or two neutrons(isotopes). Two deuterium atoms have the proper number of electrons, protons and neutrons to form helium. However, it takes energy to force atoms together, and the proper energy exchange conditions must be present in order for them to unite.

On the basis of the above analysis, one would have to conclude that there is a definite attraction between protons and neutrons, which is indeed the case. It is understandable that the two protons would repel one or the other from the nucleus since they both have positive charge, which is a plausible explanation as to why there is no tendency to form H_4. The neutrons must serve to bring the protons together in the nucleus, allowing the formation of helium from deuterium under the proper conditions.

The lack of any neutrons in the nucleus occurs only for the element hydrogen. Apparently the neutron inherently acts to attract protons, but not very many, and usually only one. There are exceptions to this rule for other elements, such as the isotopes of carbon and the heavy elements.

What is the geometry of the orbit of the electron? A currently accepted theory depicting the geometry of electron orbitals in various elements and molecules is based on various experimental data, including Nuclear Magnetic Resonance (NMR).

To briefly explain NMR, when a sample is placed in a magnetic field the nuclei tend to align with this field. A radio frequency field is also applied, and it is found that energy is absorbed at certain frequencies. This energy absorption is believed to be due to changes in orientation of the nuclei . The electrons are said to shield the nuclei, and the structure of the atoms and molecules is then interpreted from the resulting absorption data. This analysis indicates that the electrons of the hydrogen atom revolve in a spherical shell around the nucleus.

A single electron cannot, of course, form a shell, but the electrons have circular rotation, and if all of the circular paths of each molecule have no general orientation, then the measured result for many atoms and molecules will indicate a spherical shell. Also, measurement takes time, and the orientation of the circular paths can vary with time, so the data is also time averaged. Therefore, the measurement method does not provide an accurate prediction of the path of the motion of the electron.

The third element is lithium, which contains the helium structure within it plus another electron, proton and neutron. The uneven number of electrons makes it rotationally unbalanced, and it is indeed chemically active. The added electron in the lithium atom rotates in a larger outer shell which is also spherical. The inner electron shell is said to be "filled" since no more than two electrons have been found at this radius in any of the elements.

Lithium is a solid, and it is our lightest solid. It has three electrons and is therefore unbalanced. The number of protons in the nucleus determines the "atomic number," which is three for lithium. Rotationally unbalanced atoms tend to form liquids or solids due to their tendency to unite with one another (hydrogen being an exception, probably because it is so mobile). An external attractive force is exerted between atoms, and the strength of the force is proportional to its "mass." The mass of an atom is proportional to the number of protons and neutrons it contains (measurements indicate that the mass of an electron is much less than that of the proton), and the number of protons and neutrons in the nucleus is called the "mass number."

We need to say a few more words about electron energy at this point. Electron energy levels are measured in electron-Volts, but the energy level is related to the rotational energy of the electron in its orbit. The smaller the orbit, the higher its energy level. As we impart energies to the electrons, say by photon absorption, the electron shifts to a tinier, higher-energy orbit.

When an atom emits a photon (radiates), it loses energy and the radius of the orbit increases. The electron energy remaining in the atom is inversely proportional to the radius of the electron orbit. Hypothetically,the radius can reach any value and the atom will still have energy. Thus the Bohr radius of the electron orbit is limitless. In reality, however, the size of the orbit is limited due to the force required to hold the electron in orbit. Therefore, the Bohr radius is an abstract concept that somehow produces accurate prediction of the discrete radiation spectrum of certain atoms.

De Broglie suggested that matter could have a wavelike behaviour. He applied boundary limits to the waves in the hydrogen atom and obtained the Bohr quantization limits. The dynamics of wave action were established by Erwin Schrodinger who, in 1926, originated the foundation of wave mechanics. A string that is stretched between two supports can vibrate at any frequency which have multiples of a half wavelength at the lowest frequency. The strings on a guitar work in this manner, and the tones are determined by which fret is used for vibration. The action of a particle bounding between two wall leads to energy quantization by wave mechanics, similar to that of the Bohr method. Wave mechanics has been successfully applied to the analysis of the conduction of light through materials and has resulted in recent significant improvements in materials for optical applications.

What is the <u>exact</u> orbital path of the orbital electron of an atom? Has the orbit been predicted correctly by the methods described above? Current models of atomic orbitals portray paths that pass through the nucleus, while the nucleus is also regarded as a forbidden region for the electron. Thus the results are contradictory, as is Bohr's portrayal of an arbitrarily large orbital radius, and open to question. One of our goals is to determine how the orbital path relates to the forces of the atom, and possible orbital paths are examined in a later chapter.

How does matter conduct electrical current? When a photon kicks an electron out of orbit, an empty space is left in the atom. In semiconductor physics, this is termed a "hole", and any other electron passing nearby will have a tendency to fill the empty orbit. If "free electrons" are present in a material, then it is a *conductor* and electrons are therefore free to come and go if we supply the required energy to make them move. The moving electrons are called an <u>electron current</u>. In an *insulator*, the electrons from nearby atoms are tightly bound, and much more energy is required to produce conduction. When a sufficient voltage is applied such that conduction does occur, the current quickly reaches high levels, and the material is usually damaged in the process. Electrical conduction will be discussed in greater detail in a later chapter.

The orbital electron can also be forced in to a higher-energy orbit by excitation from an external radiation source. When the electron move back to its original orbit, a photon is released which produces light. In certain semiconductors, such the LED, an energy gap must be overcome before photons are produced by current flow. In gases, ionization can produce photon generation. The shape of the spectrum of radiation varies with the method by which energy is generated.

We now have a fundamental explanation of light absorption, radiation, and electric current flow, in addition to an understanding of the differences between stable and unstable molecules. These processes, are, of course, more complex than the descriptions given above. For instance, some metals conduct better than others, and no explanation was given to explain the reason. Also, only three elements were described, and we have many more to consider. We will see a pattern emerging as we the more heavy elements are formed.

Getting back to the third element, lithium, it has three electrons, three protons, and four neutrons. An extra neutron has jumped into the nucleus and likes it there. As previously stated, the mass or "weight" of atoms is primarily dependent upon the neutrons and protons since the electron has low mass (according to present theory). Hydrogen has an atomic weight of one, helium four, and now seven for lithium as determined by counting the protons and neutrons in the nucleus. The pattern that emerges for this process shows that the elements acquire an added atomic weight of either one, two, or three as the elements are arranged by their weights. Also, only one proton and one electron are added each time that we advance to an atom of higher mass.

When the masses of the elements are plotted versus the number of charges in the atom, the curve is not quite linear. The differences in incremental weight of the different elements are due to the number of added neutrons. In other words, the increase in mass with transition in elements would form a smooth curve except for the different isotopes (extra neutrons in the nucleus).

No further points to make about incremental atomic weights and isotopes since the phenomena is restricted to what happens within the nucleus of the atom and this is beyond the scope of this book. Nor will we explore any of the other internal characteristics of the nucleus where extremely powerful binding forces exist and tiny particles of various types are present. When the nucleus is blasted apart by a high-energy particle, strange results occur. Many tiny particles seem to burst forth, existing mostly for only a very short period of time. While much evidence about this phenomena has been collected, there are still deep mysteries hidden within the nucleus, and unsubstantiated theories abound. Take, for instance, the hypothetical *quarks* within the proton. While there is reason to believe that there are three quarks withing the proton, their existence is yet to be proven. The search continues, and theories are now beginning to surface to explain why none has ever been found.

Why are some materials solid, others liquid, and others gas? The noble elements, such as helium and neon, are mechanically and chemically stable, having little tendency to combine with other atoms of any type. It takes much force to make them into a liquid at room temperature. If the temperature is low enough, and if the pressure is high enough, then they, too, can become solid. Unbalance atoms have a greater tendency to form liquids and solids.

Lithium is an unbalanced atom since it only has one electron in its second orbit, which is larger in radius than the inner orbit, and therefore combines with another atom. This larger shell is also believed to be spherical. This atom must also be unstable since it attracts another atom. The result is a lightly-bound mass or what we call a "solid."

But what about liquids? Why isn't lithium a liquid? Well most solids do form liquids as the temperature is increased, but it just so happens that lithium is a solid at room temperature. In the liquid phase, the intermolecular binding energy must be greater than that of the gas phase, and in the solid phase greater still. We said that lithium is lightly-bound, which indicates that the molecules are fairly stable, even though the atoms by themselves are not. Lithium is light in weight (it floats in water) and can be cut with a knife. Both lithium and hydrogen can react with other elements, and this is characteristic of elements with only one electron in a given shell (or unfilled shells).

As we add more electrons, protons, and neutrons to form new elements, not all of the orbitals are in the form of spherical shells. The next element is boron, and its outermost electron shell is not in the form of a sphere but appears as a balloon pinched in the center. As atoms acquire a greater the number of electrons, protons, and neutrons, the number and type of energy shells increase. The mechanical strength of <u>solids</u> that are composed of simple elements is generally stronger for the heavier atoms (until they get too heavy and start to come apart). The strongest materials do <u>not</u> have filled energy shells and may have an odd number of electrons one of its shells. Fully balanced atoms exist primarily in the form of gases and are called *noble elements* since they are chemically inert. The description of the details of all of the atoms are given in texts on chemistry (see the bibliography).

    The basic method of the construction of atoms has been described. Whether an atom is metallic in nature, an insulator, a heat conductor, has chemical activity, or is radioactive is dependent on the electron/proton/neutron configuration. We will be concentrating on the basic characteristics of the hydrogen atom, since it is the most simple element. Once the internal properties of the hydrogen atom are understood, theories can be formulated, and these theories can then be extended to the heavier atoms.

*The question whether space is real
apart from space-filling objects . . .
dates from early times.*

---W. T. Harris

CHAPTER VI

A Universal Force

Is there a "Universal Force?" Scientists have long been trying to identify a universal force, but all theories have thus far been incomplete. There are four basic forces in nature that have been identified: *gravity*, *electromagnetic*, *weak*, and the *strong* forces. Certain eminent scientists, such as Einstein, have spent many years trying to combine these forces in a single unified field theory and failed. If there is a universal force that produces these four manifestations of force, then how does it occur? The attempt to identify the universal force of the universe will be made, beginning with this chapter.

Before we begin, we should understand what constitutes a *field*. We are immersed in fields of various types. The gravity of the earth produces a force field that holds us to its surface. The further above the surface of the earth, the lower the force of attraction. Another gravitational force field comes from the sun, which holds the earth in its orbit. The terminals of the battery in a car has a tiny electrical force field surrounding it. The electrical wiring in your house forms magnetic fields due to the current that is flowing through them. The power lines produce much stronger electrical fields. Every atom in the universe produces force fields, and the nearer they are, the higher the force field.

Let us first look at the four basic forces of nature. The first force that was recognized is the force of *gravity*. Gravity is by far the weakest force in the universe; so weak in comparison with the strong force that the amount of difference challenges imagination. Newton's theory of gravity has describes it quite well, at least up to a point.

The *electromagnetic force* is the attraction or repulsion of electrical charges. An example of this comparatively weak force is the buildup of static electricity. Pulling a sweater over your head on a dry day, you can feel the force on your hair which is produced by static electricity. Magnetism is another form of an electrical field. Maxwell's equations define the relationships of these forces.

The strongest forces, however, exist <u>within the atom</u>. Therein lies what are known as the *strong force*, and the *weak force*. The strong force is the force that holds the nucleus of the atom together. It is by far the strongest of all the known forces. It takes a shot from a high-energy atomic particle to blast the nucleus of the atom apart. The weak force has to do with the energy generated by radioactivity, the unstable nuclei of large atoms radiate energy as they radiate protons and neutrons.

The strong force is 100,000 times stronger than the weak force and 137 times as strong as the electromagnetic force. The electrical force is about 10^{39} times as strong as the gravitational force. Take the gravitational force and multiply it by ten, and then do this thirty-nine times (a ratio of a thousand billion, billion, billion, billion). In other words, gravity is certainly not a very strong force.

If there is a universal force, then these four forces must somehow be tied together. Einstein equated inertial and gravitational forces, but the difference between gravitational force and atomic force is so enormous that he was unable to derive the "structure laws" that combined these forces together, and he stated that "it belongs to the future" (to solve this problem). His two realities were *field* and *matter*.

There have been more recent efforts at unifying these four forces. There are several related theories that have a degree of acceptance: "gauge theory," "electroweak unification," "quantum chromodynamics" (QCD), "quantum electrodynamics" (QED), and "grand unification theory" (GUT).

Gauge theory has some aspects that will fit the models of this book. Gauge theory advances the proposition that measurements of a given type can be made at various points in the same manner with the same result. An example would be to take a ruler and measure the length of an object. If the result does not change under various conditions, then gauge theory applies to those conditions. By the Theory of Relativity, spatial coordinates change with velocity, so gauge theory does not apply when measuring the length of an object that is moving with respect to the observer. The variation in length with velocity is, however, common to all systems from a point of reference within the system, and gauge theory does apply to this relationship. In other words, the length of an object doesn't change within moving systems, but it does change with respect two systems moving at different speeds.

Another concept is that of symmetry. An example of symmetry is a circle (symmetric in a plane), or a sphere, which is symmetric in three dimensions. Gauge symmetry is either local or global. Local symmetry applies only to a local area, while global symmetry must hold everywhere. Electricity exhibits global gauge symmetry. Electrical voltages are measured with respect to a common point called ground. Global symmetry requires a single reference point, while local symmetry allows any local reference point from which to make measurements.

In local gauge symmetry, the definitions of one observer do not depend upon the definitions of another. This requirement is fulfilled whenever two effects can cancel each other out. In other words, the system is allowed to change in such a way that two forces cancel out, leaving a net force of zero. In this way, two or more great forces can produce a weak force. Gauge theory can be applied to the exchange of energy between atomic particles. Electricity and magnetism are said to be different aspects of a single force because they both exchange photons. We will investigate these forces further in a later chapter, although the techniques of gauge theory will not be employed.

In the *electroweak* unification theory, electromagnetism and the actions caused by the weak force, which are produced by a family of particles, are different aspects of a single force. The electroweak theory is supported by another branch of physics, QED. When all of the possible interactions between various atomic particles are considered, they can be interconnected such that a several strong forces can produce a much weaker force. The particles involved are neutrinos, mesons, leptons, and bosons. It will be shown that particle physics is not essential to the basic theory of gravity.

Similarly, the *QCD* theory will only be discussed briefly. The nucleus of the atom is believed to contain particles called *quarks*. Quarks are said to have fractional electrical charges, and colors have been assigned to them as a simplified way to represent their charge.

These quarks are said to exchange *gluons* which makes them change color. A quark have never been identified in the laboratory, and the proton has never been observed to decay into quarks. Quarks are also used to explain the rejection of the electron as it approaches the nucleus.

The GUT theory tries to tie these sub-theories together by a family of twenty-four massless particles. The symmetry is broken in collision between high-energy particles, and the electroweak force is exhibited. For lower energy collisions, the three forces, strong, electroweak, and electromagnetic are all present. A main result of the GUT theory is that protons are predicted to decay into a positron and a meson which in turn decays into a pair of photons. Such an event has never been witnessed. However, the decay time is an extremely long period of time, and perhaps no one has been around long enough to see the event. The theory model can easily be changed to accommodate a shorter period of time if one is ever seen.

So what ever happened to gravity in all of these theories? There is no known cause of the force of gravity; it is simply assigned to a characteristic of mass. It isn't there yet; an accepted theory, that is. That is the job of the current batch of scientists. They are supposed to work with new inventions (of theoretical scientists) called the *graviton* and the *gravitino*. The gravitational force between the earth and the moon is believed to be produced the flow of small particles called gravitons. In violent collisions, gravitino exchange predominates and modifies the gravitational attraction. These scientists are expected to tie the actions of the gravitino in with the GUT. The resulting forces exhibit *supersymmetry*, and the quantum theories of gravity using supersymmetry are called *supergravity*. None of these imaginary particles have ever been found in any experiment.

It all seems a bit complex and surreal, doesn't it? Wouldn't it also be nice to have scientific theories that are compatible with realistic concepts? A universe that is ultra simple? Scientists have thought of that too, and it is called *superunification*. There will supposedly be only one basic particle, the *superparticle*. The problem is that they may be going in the wrong direction since many more new particles have been predicted by these theories. At any rate, it is a *great idea* (GI).

Physicists devise experiments and use scientific measurements to develop the models that permit prediction of the action and reaction of matter and fields. We have seen that these models are not all compatible with one another, nor do they all have much reality in their conception. They would like to build a simple universe, but they seem to be digressing, since complexities lead to more complexities. The goal, however, is admirable.

The previous chapters were intended to provide some perception of the present and past conception of scientific thought, as it has now evolved. From this information we can begin to build a universal theory which is compatible with electricity, force, motion, and what is considered to be mass. A realistic model of matter is desired. In order to do this we must, at least temporarily, discard some of the old (and some of the new) theories and proceed with new lines of thought.

Another question, that has been around for quite some time: "Is there an ether?" The possible existence of a medium that allows the propagation of energy through space has bothered many scientists over the years. If there is but a single universal force, and if there is an ether, then they must surely be related to one another. A 19th century idea envisioned a "luminiferous" (light-bearing) ether which served to transport light through otherwise empty space by means of optical vibrations. A magnetic somehow penetrates a vacuum, and if the field varies with time it is called a field. But then what is it that carries this field from one point to another? Water is a medium that produces mechanical waves, so what is the medium in space? Two scientists (Michelson and Morley) conducted an experiment in an effort to test the ether theory. They beamed a ray of light over a long path, at the end of which there was on optical reflector that sent the beam back over the same path. The absence of a Doppler effect pretty well destroyed the belief in the existence of an ether. Is the idea of an ether really that unbelievable? While not many scientists believe in it, any physics texts at least discuss the subject and cannot totally disprove it in spite of the above mentioned experiment. While the idea of an ether is not essential to the arguments that will be presented, it is nevertheless a possibility to be considered and not merely dismissed.

The following hypothesis will now be proposed:

A BASIC TENET "A UNIVERSAL FORCE":

All mass in the universe is *electrical* in nature. All of the forces that are in existence are forms of electrical forces. Time and space are not electrical, but these are "dimensions" which are the independent variables and are part of the definition of electricity.

We begin by accepting the possibility that a universal force exists and that it is <u>electrical</u> in nature. Now we can kick at it a bit and see if it holds up. If we accept the idea of a universal electrical force, then we can say that there are five dimensions: the three dimensions of space, time, and the positive and negative electrical charges. Then what about the neutron? Is it not electrically neutral? Well, yes and no. Basically the answer is yes, it is electrical but not very active, electrically. Since it exerts force and everything is electrical, then we must conclude that it, too, must be electrical. And the neutrino? Yes, it must also be an electrical wave. The main stumbling block has always been the gravitational force, and the emphasis will be on solving the mystery of how it is produced from electricity. Once the mystery has been solved, additional inferences about the above mass and force fields will follow.

In the previous chapter, we saw that all mass is built from atoms and molecules. These tiny building blocks consist of electrons, protons, and neutrons. The subatomic particles are still being investigated at this time, with physical models of their characteristics being formulated. Many new theories have been advanced which use these smaller particles to account for the four forces of the universe that were described above. The strong force is attributed to the energy required to confined the protons and neutrons within the nucleus of the atom. Since electrical sub-charges, called *quarks*, are believed to exist within the proton, the quark may be functional to the strong force of the universe. Subatomic particles are not an essential part of the arguments of the following chapters.

The neutron presents a bit of a problem since it is electrically neutral. It is believed to consist of an electron and a proton, and yet it exhibits the characteristics of mass. But then, so does hydrogen, and it also contains an electron and a proton. The assertion is, that if it can be proven that the gravitational force affecting the hydrogen atom can be related to the electrical force, then a similar condition will occur for the neutron. Some evidence will be presented to support this idea.

If all that we had to consider was the attraction of the electron to the proton, it would be easier to conclude that the entire universe is electrical in nature. However there are two serious dilemmas which would frustrate us in reaching this conclusion:

>Dilemma 1: Electrons, protons, and neutrons are themselves considered as *masses* due to their attractions to electrically neutral bodies. The relationship between mass and electricity has never been determined.
>
>Dilemma 2: Our current model of electricity does not fit in with the mathematical formulas for gravity and attraction between masses. When electrical formulas are applied to electrons and protons, the resulting forces are <u>far greater than gravitational forces</u> and do not fit the model.

These are knotty problems that Einstein struggled with in his search for a unified field theory. If these dilemmas can be resolved, then the idea of a universal force, or universal field can made plausible. Resolving these dilemmas will be two of the major tasks ahead of us.

Consider dilemma number one. Within an atom the electron is considered as a <u>body</u> rotating around the nucleus. This electron is an electrical point charge which has an attractive *force* between it and a proton (Coulomb's Law: $F=q^2/4\pi r^2$, where q is the electrical charge and r is the separation between charges). But this force field is really an electrical field (since the electrical field strength is determined by the force per electrical charge, $E=F/q$) in the form of an electrical dipole as picture in Figure VI-1. In this figure, the lines of threes sets of curves portray the spatial field distribution at three different levels of force. The hydrogen atom can be represented by such a dipolar electrical field rotating about the positive point of the dipole. It is also a rotating force by virtue of the field equation.

Recalling the Rutherford picture of the atom from Chapter V, the electron and protons are depicted as points in space, and the illustration of the hydrogen atom as a rotating dipolar electric field represents it more accurately. The concept of mass provides a method by which forces acting on atoms are metered. Dilemma number one can

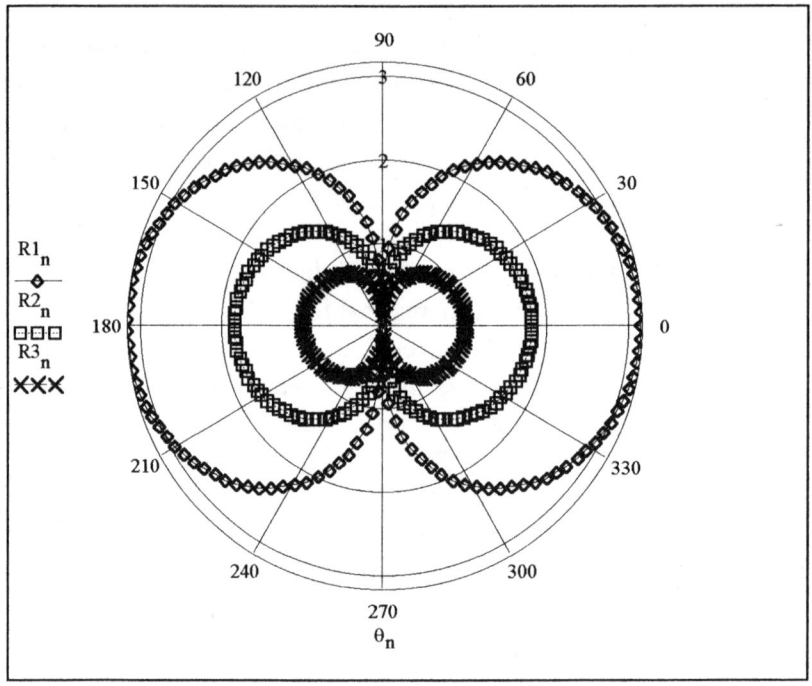

Figure VI-1. Plots of the Field Strength of a Dipole
at Three Intensity Levels

be resolved by accounting for the *electrical forces acting on the electrons and protons*, and our concept of mass is reduced to a concept of electrical force fields. That electrical forces are produced *within the atom* is fairly well accepted by the scientific community (the weak force), and the force levels agree with measurements.

But then what about *outside the atom* where there is attraction of an atom to a neutron? Why didn't scientists simply use the concept of electrical force fields rather than stick with the concept of mass as an entity by itself? There are several apparent reasons. The neutron is believed to be electrically neutral, and it contributes to the atomic weight of an atom. The forces of matter were attributed directly to the mass concept a long time before electricity was ever discovered, and old ideas are slow to disappear. After electricity was discovered, the electrical laws show that two equal and opposite charges cancel, and electrical cancellation is believed to occur outside of the atom. It was concluded that, due to these factors, it could not be an electrical effect.

With the gravitational theory of Newton and the electrical field equations of Maxwell, we still have dilemma number two, which implies that an enormous difference electrical and gravitational forces exists external to the atom. This reasoning has led to the belief that electrical force and gravitational force are two separate phenomena. The energy within the atom dwarfs the energy outside the atom, and this difference is much too high to be accepted as the common denominator of the gravity equation. Einstein suggested "modifying our field laws in such a way that they would be valid everywhere, even in regions where energy is enormously concentrated." He was unsuccessful and accepted the ideas of two types of forces.

For the moment, bear with me and presume that such solutions do indeed exist, and that the gravitational force is another previously unknown manifestation of electrical forces. Avoiding the consideration of mass as a constant results in an entirely different vision of matter that will be shown to consist wholly of electrical forces. In order to picture the electrical force within the atom better in one's mind, consider an analogy with a tornado. The tornado has very fast and powerful winds revolving about a tiny center which is very calm in its center. The electron revolving about the proton presents an analogy to the tornado with the winds replaced by a revolving electrical field. Thus, with this concept, the usual idea of "mass" of the electron and proton is replaced by the interaction of electrical force fields. Hence, in the forthcoming investigations, the usual concept of mass as a constant must be abandoned in favor of a dynamic force fields. Therefore, the mass constant must be used with caution in any equation of physics due to this limitation. The evidence to support this conclusion will be presented later.

If we stop to reflect on the above picture of atomic action, we can visualize something nearby, say a wall, as an immense cluster of tiny, but powerful, whirling force fields acting together, so small that they seem to be particles ("force dots"). We, ourselves, are just more of the same type of conglomeration of little, strong whirling forces. If there are forces in the universe that can pass through these little whirlwinds without disturbing them (and we have good reason to believe that there are), then our little world is almost transparent to these types of forces (high-energy particles).

Then what is the difference between a particle and a force field? It is primarily a geometric problem. An electrical field extends out into space an immeasurable distance. Its strength diminishes rapidly with distance but continues to the depths of the universe. The atom is a group of these forces fields, and the atom is presently considered as a particle. Using this definition, a particle is the center of a rotating force field (or a group of force fields with a common center).

Another way to view the electrical forces of the atom is to picture each electron and each proton, anywhere in space, as being connected by an invisible rubber band. The universe, by this portrayal, is interconnected with a huge web of imaginary rubber bands that keeps the universe together. Since each atomic dipole rotates, the electrical fields become waves that are in constant rotational motion. The universe is a sea of electrical waves, in continual motion around us, and all moving at extremely high speeds in various directions. While the field of each individual atom is very weak, the sum of the forces of all of the atoms control every event in the universe.

We are all very familiar with the current mass concept, and it is difficult to abandon old ideas. The above image of the universe, as a sea of electrical fields, is difficult to visualize since we cannot see electrical force fields, and the winds of a tornado are not exactly the same as an electrical force field. We use our human senses in perception, and they are very limited. Our present model of the universe took many centuries to develop, and the diverse tools for measuring the universe have progressed rapidly in recent years. But progress has been more rapid in certain specific areas, leaving huge gaps in our knowledge of other areas. We relate more easily to what our body senses tell us, and the ideas of the past linger on, thereby producing a gross distortion of the picture that actually exists.

Dilemma 1 can therefore be removed by simply regarding mass as the electrical characteristics of atoms. No one is expected to have much faith in this assumption at this time, since no proofs have yet been offered to substantiate it. Making the assertion is simply the first step that is required to establish the concept of a universal field. The forces acting on all masses, are therefore produced, in some way by, electrical forces, which is the reason that equations containing mass constants must be carefully scrutinized.

The next step is to account for the very weak gravitational force in terms of electrical forces. No one has ever been able to accomplish this task, so do not expect it to be easy. Once this is achieved, the idea of a true universal force becomes much more believable.

Electricity--- greatest servant of man, itself unknown.

---C. W. Eliot

The more a man looks at a thing,
the less he can see it,
and the more a man learns a thing,
the less he knows it.

---Chesterton

CHAPTER VII

Resolving the Gravitational Force Dilemma
A New Approach to the Problem

In the previous chapter, the assertion was made that there is only a single force in the universe. However, two difficult dilemmas were uncovered. Dilemma number one, the relationship between mass and the electrical particles, was resolved by the assertion that the only type of force in the universe is electrical, and therefore the forces acting on a mass are a consequence of electrical fields. Einstein was unable to accept this premise since he was unable to solve the problem of a weak gravitational force produced by the very strong electrical field. How a gravitational force that is so very weak could be produced by an electrical force that is so strong is dilemma number two. Now we need to investigate the process by which the very weak gravitational force is produced from the extremely strong electrical forces within atoms. Once this problem is solved, both dilemmas will be removed, and the assertions of the previous chapter will then be substantiated by the new evidence.

Matter has been pictured as being essentially empty except for immense clusters of tiny whirling electrical force fields. Two separated electrical charges form an electrical field, and this electrical field corresponds to a force field. However, the difference in strength between electrical and gravitational forces is so great that it is difficult to even imagine. We are going to attempt to resolve this dilemma, but it cannot be done by merely substituting constants into simple formulas.

If we assume that all fields are electrical, then the use of the mass constant in physical equations must be utilized with care. The reason for exercising caution will become evident after the analysis begins. This restriction does not permit equations to be manipulated in the normal manner, and different approaches must be adopted.

All of the constituents of the earth are formed from molecules which contain atoms of all of the elements (except for a few man-made ones that are unstable). Each atom is an ordered group of electrons, protons and neutrons. Disregarding the smaller particles, all of the universe may consist of just these three particles (plus the tinier subatomic particles within the nucleus that will be neglected for the moment). As was stated earlier, each atom has a proton for each electron, plus neutrons (except for hydrogen which has no neutron) which are electrically neutral. Because two equal and opposite charges cancel, each of the atoms is believed to be electrically neutral and the single electron should therefore have no attraction for any atom. Because the electron has a measured weight or force acting between it and the earth, and, since other masses attract each other, there must therefore be some other explanation of the phenomenon.

Even with the assumption that it must be an electrical attraction of some sort, the answer is far from obvious. So what might cause these forces of attraction? The electron is said to be attracted to the proton in the nucleus, but it cannot approach too close to it because of an invisible barrier. At a far distance from an atom, no electrically attractive force is believed to exist, and yet a force of some type is present which has been assigned to a special characteristic of its "mass." Evidently something very unusual has been either overlooked or never found by anyone up to this point.

When once you have taken the impossible into your calculations, its possibilities become practically limitless.
--- Saki

Ch. VII - 95

We will be looking for a secondary effect, <u>something that everyone has overlooked up to this point</u>. Because of the greater force between electrical charges, an unbalance of the electrical field of the atom must exist. Because the strength of the electrical forces is so high, it will not take much of an unbalance to produce an external force. We will be looking for an unbalance which is so slight that even a powerful computer will have a difficult time performing the calculations, since the numbers of our calculations will stretch out to many decimal places (sixty-three). In engineering, when unbalance occurs, we look for a *nonlinear effect*. The assertion has been made that the forces from the electrical charges do not quite cancel outside of the atom. Therefore we should look for a nonlinear process wherein cancellation is not complete. To put it in simple terms, we are dealing with a highly balanced atomic force system which has a region where addition and subtraction are not quite exact.

In most engineering applications, the situation is usually the other way around. Every system is nonlinear to some degree, and a region of linearity is desired, where distortion is minimal, so that linear operation can be obtained with minimal error. A transistor, for example, is very nonlinear, and transistors and transistor circuits must be carefully designed in order to obtain linear operation. In another example of nonlinearity, when a spring is pulled far enough apart, the force exerted is no longer proportional to distance. The problem, however, is not that easy. We are looking for a needle in a very big hay stack. If a computer solution is sought, then the "delta" error of its mathematical algorithm can far exceed the small numbers in our model.

So what, then, is to be the approach to the problem? Quantum mechanics cannot be of much use in this situation, since it is based on energy concepts and probability theory, and we must deal with force equations which do not necessarily depend upon energy or probabilities. Classical analysis, on the other hand, can provide the necessary tools since the *force equations* can be manipulated in various ways. The area of classical analysis that we will employ is *nonlinear analysis*. Unfortunately, nonlinear analysis can be very complex and easy solutions are rare. A nonlinear effect, however, is essential to the solution that we seek, and we know that it must be very slight.

Therefore, we will begin with the nonlinear equation that

describes how the electrical force field varies with distance. Furthermore, the gravitational effect must be a result of the <u>external</u> electrical field produced by the electrons whirling around the nucleus of the hydrogen atom. The Rutherford model of the hydrogen atom fulfills these requirements.

The Rutherford model of the atom presents additional difficulties that must also be resolved:

- ♦ A planar orbit does not produce isosymmetry. In other words, the attractive force is expected to vary with the orientation of the orbit, and we know that gravity does not have this characteristic. The requirement for isosymmetry may dictate another form of the electron orbit.

- ♦ The accepted model of the hydrogen atom has an electron rotating around a proton. The two charges form a rotating dipole whose field reaches far out into space. The characteristics of the rotating field presents some new questions about the <u>form</u> of this rotating field.

So we have two new and difficulties to deal with. These new perplexities must also be untangled in the process of solving the problem.

The approach to be taken has now been outlined. The premise is that the force of gravity is produced by a consequence of the electrical field, it must be a nonlinear effect, and it must result in a very slight unbalance of the external electrical forces. We will next analyze the dynamics of the field of the hydrogen atom.

Nothing comes easy.

CHAPTER VIII

The Rotating Dipole of the Hydrogen Atom

In the previous chapter, the approach to solving the problem was defined. The assertion is that a *nonlinear effect* of the electrical fields within the atom produces the weak external gravitational force. We have chosen the Rutherford model of the hydrogen atom for our analysis, and it consists of a rotating dipole (an electron and a proton, which are separated in space, form an *electrical dipole*) wherein the electron rotates about the proton. We will now consider the field characteristics of a rotating dipole.

In the earlier chapters, each time a solution to a problem was proposed new difficulties surfaced. The new enigmas revealed by this chapter are particularly imposing. The analysis that you are about to witness produces results that exercise some aspects of the Theory of Relativity.

It is an exercise of the imagination to envision the structure of the field of an atom. The electron and proton are not particles as they are often depicted, and so the illustration of Rutherford's atom, in Figure III-4, only represents the two dipolar centers of the field, the electron and the proton. Also, the electrical fields within atoms are tremendously strong compared to the external gravitational field. These fields decay rapidly with distance (inversely with the square of the distance, termed "inverse square law"). Within the hydrogen atom there is but one electron and one proton (the possible presence of an additional neutron will be neglected in this discussion). In the Rutherford model, the electron is separated from the proton and is rotating about it. A separated electron and proton form a dipole, which in the case of the hydrogen atom is rotating asymmetrically about its geometric center.

How fast does the electron rotate about the proton of the hydrogen atom? Present estimates indicate that it is probably several times slower than the speed of light. Since there is a fair degree of uncertainty in this assumption, let us contemplate the dynamics of the

Ch. VIII - 98

rotating dipole.

Light travels much slower in materials than in a vacuum, which is an indication that it is delayed by some sort of interaction with the electrical fields of atoms that it encounters. The differences in the speed at which light travels causes a light beam to bend, and these characteristics are utilized in the design of imaging lenses for cameras and telescopes. The speed of light, for some reason, is also limited in passing through a vacuum. This peculiarity has led many scientists to believe that an *ether* exists in the universe. The idea of an ether has intrigued scientists for many years. Einstein called it the "ether-sea" and discussed the difficulties in proving that an ether exists. Modern physics books may still mention the ether, but it no long has the attention that it once had.

Then what is it that limits the speed at which light is transmitted through space? A measure of the degree of optical time delay in a material is given by the dielectric constant, ε_0, which represents the static electrical field strength. The strength of the magnetic field is represented by the permeability constant, μ_0 (which we will assume to be due to the rotating electrical field). It has been determined that the speed of light in a vacuum is related to these parameters by the relationship (from Maxwell's equations).

$$c^2 = 1/\mu_0\varepsilon_0 \qquad \text{[VIII-1]}$$

This equation has <u>two solutions</u> which differ in sign. The right-hand side of the equation consists of two electrical constants that are used to define and scale the electrical parameters inductance and capacitance of an electrical circuit. But equation [VIII-1] applies to plane waves. The following slight modification of the above equation is chosen for reasons that will become apparent.

$$c^2 = (-)1/\mu_0\varepsilon_0 \qquad \text{[VIII-2]}$$

There are also two solutions to the above equation, resulting in vectors with equal magnitude (speed) but of opposite directions, as was the case with equation [VIII-1]. However, the new speed vectors are orthogonal (at right angles) to those of the first equation. Speed is the rate of change of distance with respect to time. If the speed is considered to be a derivative, as in calculus, then the equation becomes a second-order differential equation whose solution represents the frequency of an oscillator ($\omega^2 = 1/LC$). The frequency of this oscillation is the square root of the right-hand term above. This proposition results in the implication that a photon may be a revolving electrical field in space, moving at the speed of light (plane waves are not know to exist in nature). There are, however, new complications that must be examined.

The photon is believed to be generated by a change in the orbit of an electron about the proton. If the photon moves at the speed of light, then is the electron also moving at the speed of light (or faster)? If it is moving slower than the speed of light, then how does it generate a wave that is traveling faster? These are tough questions, and the answers are not obvious.

I am not aware of any published report of a detailed effort to analyze the dynamics of the field of a rotating dipole. Remember that the rotating dipole generates a moving field in space (a wave). The characteristics of this moving field can take various possible shapes, which creates some new enigmas as will be shown. The Rutherford model of the hydrogen atom with the electron rotating about the proton will now be analyzed as a rotating electric dipole.

The spatial field around a dipole was pictured in Figure VI-1. The electrical potential of a dipole varies in level with the inverse of the square of the distance from the center of the dipole and extends outward from the dipole to any chosen distance. The field is vectorial; that is, it varies in strength with the angle about the dipole. Therefore, since the electron is revolving about the proton, the dipole is rotating at the same angular rate.

The rotational speed of the moving field of a dipole is given by

$$\upsilon = \omega r = \text{velocity of the moving electric field}$$
at a radial distance, r.

Evidence will be presented which indicates that the velocity of the rotating field may be quite high, higher than previously thought. If a

Ch. VIII - 100

very conservative value is chosen for the rotational rate, say

$$\omega = 10^{10} \text{ radians per second,}$$

and at a remote point

$$r = 3 \times 10^{-2} \text{ meter}$$

then

$$\upsilon = 3 \times 10^{8} \text{ meters/second,}$$

and the rotational field wavefront is moving at or near the speed of light at a distance of about one inch from the little atomic dipole. The radial distance at which the field travels in a rotary direction decreases further for much faster rotations. In this example when the radial distance is greater than an inch or so the field will be moving in a circular direction at speeds exceeding the speed of light!

The fact that the speed of a rotating field exceeds the speed of light (in a *circular* direction) does not necessarily provide a contradiction to Einstein's theory, which applies to moving objects and radiation (Einstein may not have even considered a non-radiating and rotating field). A rotating dipole is not known to be a radiator of energy, and the Theory of Relativity applies to the outward propagation of energy. The question is whether or not the *rotational* speed of the revolving wavefront, generated by a rotating dipole, can exceed the speed of light. It is not possible to have propagation if the field does not radiate energy. It is also possible that the electrical field is somehow distorted during rotation, such that the rotary field movement is also limited to the speed of light.

If the speed of the rotating wave is limited to Einstein's theoretical maximum speed, the frequency of the electrical field must be inversely proportional to the distance from the center of the dipole at radial distances exceeding 0.03 meter (for the above example). Can it be possible for the frequency of the rotating wave to vary with distance? This possibility is highly unlikely. Even if the slower electron speed of current estimates is correct, the radial distance at which the speed of light is exceed is simply increased to 0.2 meter (eight inches). Either there is a limit to the speed of all electrical fields, in which case electrical fields must somehow be distorted as a function of speed, or else there is an exception to the speed limit of certain electrical fields which are not radiative. Quite a thought-provoking dilemma!

> *CONCEPT #1:*
> One possibility is to assume that Einstein's speed-of-light theory <u>does</u> apply to a rotational electrical field, even if the field <u>does not</u> radiate energy.

If this concept is indeed true, then the rotating dipolar field must necessarily be distorted, with rotation, such that no parts of the rotating field exceed the speed of light in <u>any</u> direction. A plot of the speed of a point in the electrical field then appears as in Figure VIII-1, approaching the speed of light but never exceeding it. The rotational frequency decreases with distance as shown in Figure VIII-2.

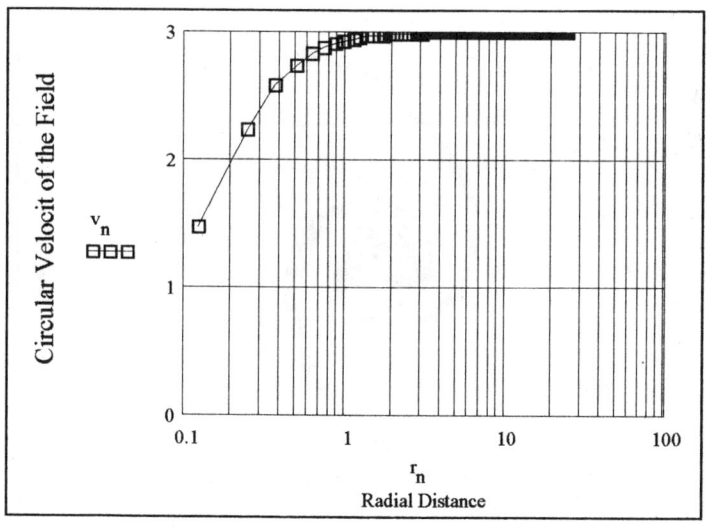

Figure VIII-1. Rotary Velocity of the Electric Field of a Dipole as a Function of Distance

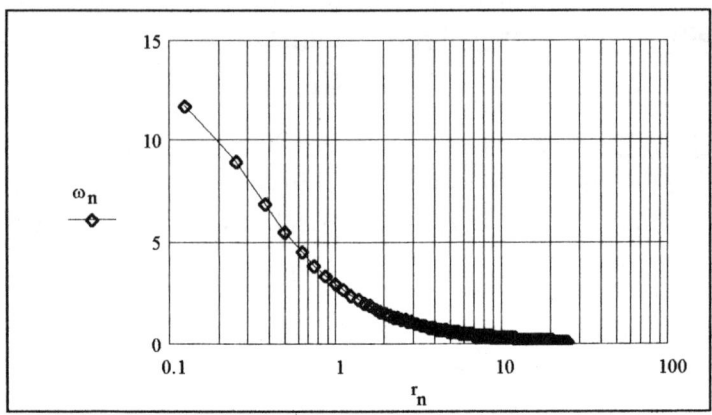

Figure VIII-2. The Frequency of Rotation of Points in the Field as a Function of the Radius

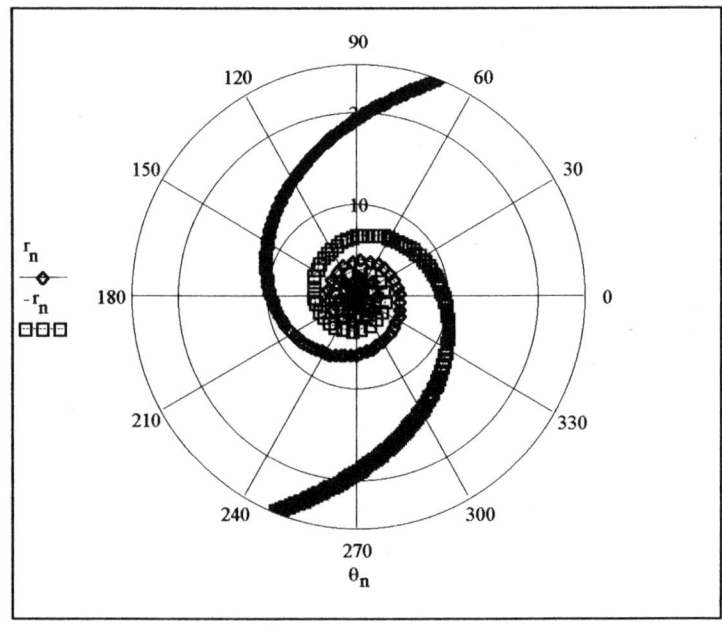

Figure VIII-3. Distortion of the Axis of the Field of the Field of a Rotating Dipole for Concept #1

Ch. VIII - 103

The distortion of the constant field-strength ray, for this example, is shown in Figure VIII-3, above. For Concept #1, the ray representing the axis of the stationary dipole becomes distorted in the form of spiral as the rotating dipole rotates. It is observed that the frequency of rotation slows as the distance from the center of rotation is increased. This result indicates that the remote forces, exerted by the rotating field of an atom, are of <u>lower frequency</u> at more remote distances. Thus the spiral changes shape with time, winding up as a clock spring, as indicated in Figure VIII-4.

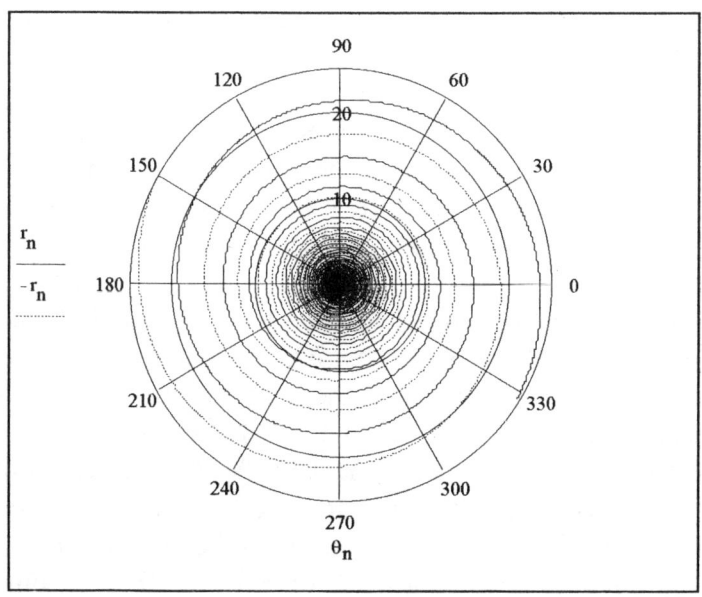

Figure VIII-4. The Shape of the Field of the Velocity-Limited Rotating Dipole Changes With Time

These results are hard to believe, since there is no evidence, in the form of measured data, to support them. A field that changes its frequency of rotation with distance would be expected to also exhibit changes energy and frequency with time, and we know that this does not happen. If we are to hold to the idea that <u>nothing</u> in the universe can travel faster than the speed of light, then either Concept #1 must be correct, or else the Rutherford model of the atom is wrong. Neither of these conclusions is easy to accept.

CONCEPT #2:
The counter-theory is that the rotating (atomic) dipolar fields can indeed move circularly through space at speeds <u>exceeding the speed of light</u>.

Although this concept is also very difficult to accept, it is somewhat defensible. When there is no energy radiation from the hydrogen atom, it is not possible to sense the velocity of propagation, since propagation does not occur. In other words, no energy transfer exists for this type of rotating field, whereas light is the transmission of energy in an optical form. Therefore, if concept #2 holds true, then we are not violating any prior basic concepts regarding the radiation of energy.

As with the Concept #1, we have no measured evidence to support or deny the validity of Concept #2, and therefore it, too, is very difficult to prove or disprove with the evidence at hand. The major arguments against it is that it would not <u>seem</u> to conform to Einstein's theory regarding the speed of a moving field. Perhaps Einstein's theory, regarding the speed of light, may not apply to the circular movement of a field. The field strength diminishes with the area covered, which correlates with the "spreading loss" of radiation that is commonly measured in antenna systems.

There is a third concept which can be used to explain the limitation of the speed of light:

CONCEPT #3:
The dipolar fields are distorted such that the field level radials are curved in such a way that the *differential radial velocity* conforms to the speed of light.

This concept makes more sense than the first concept, since the frequency of rotation is constant for any point on the wavefront. However, the <u>rotational speed</u> of a point on the wavefront of the field would still be traveling faster than the speed of light beyond a given radius. In other words, the rotational speed can exceed the speed of light beyond a certain distance from the center of rotation, but the wavefront is delayed with radial distance such that the advance of a point along the wavefront occurs at the speed of light <u>in the radial direction</u>. This field is pictured in Figure VIII-5. The curve changes shape with distance as show in Figure VIII-6, and the radial speed increases without limit as in Figure VIII-7. The shape of the curve becomes more pronounced with distance as shown in Figured VIII-8.

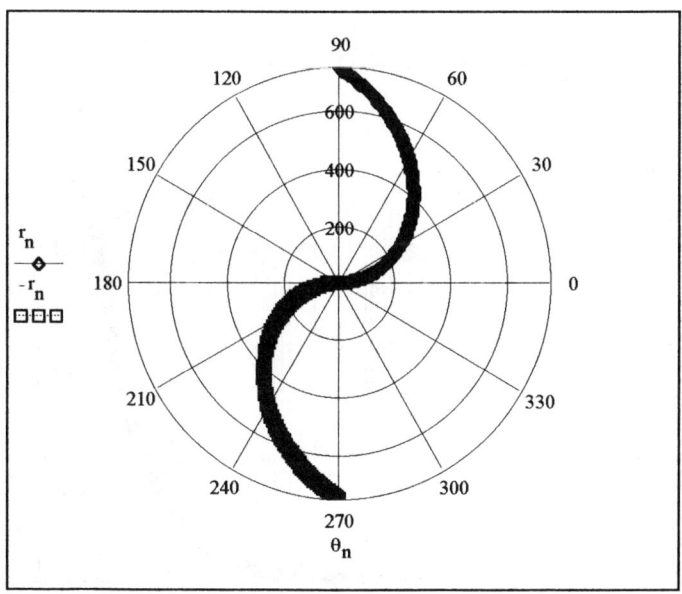

Figure VIII-5. Rotating Field of the Hydrogen Atom With Radial Time Delay

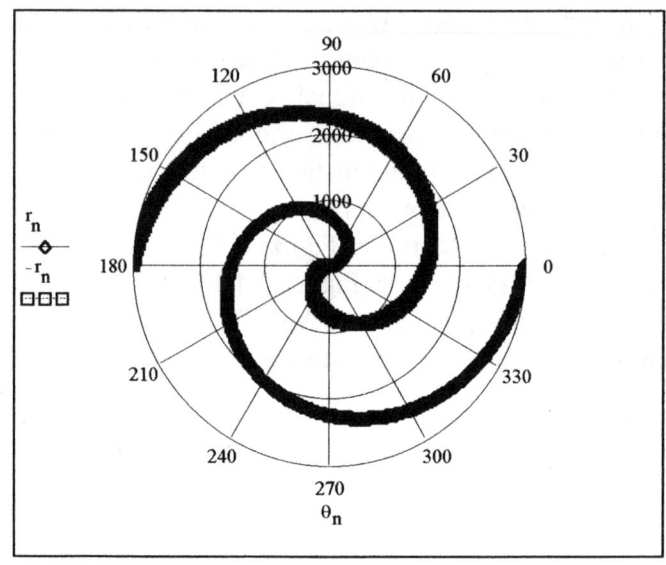

Figure VIII-6. Rotating Field of the Hydrogen Atom for Radial Time Delay At Short Distances

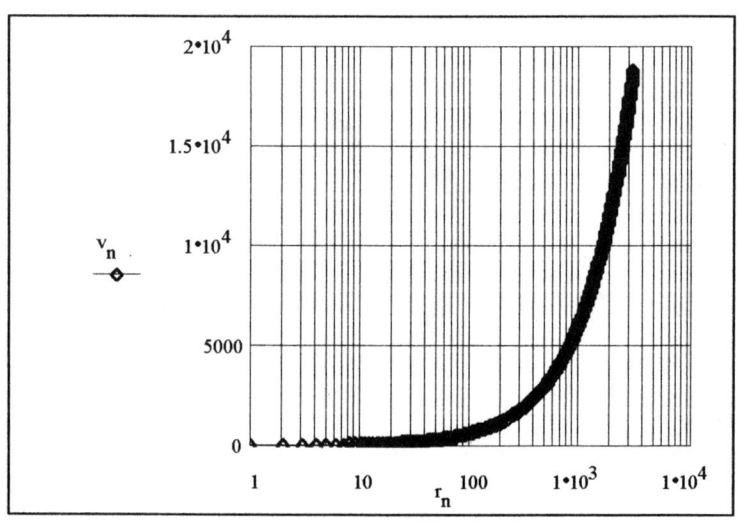

Figure VIII-7. Rotary Velocity of the Field of the Hydrogen Atom For Radial Time Delay

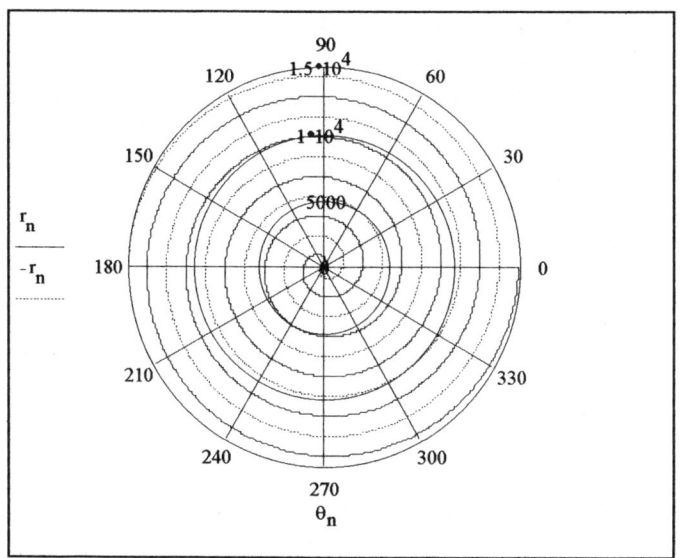

Figure VIII-8. Rotating Field of the Hydrogen Atom For
Radial Time Delay At Greater Distances

In order to verify Concept #1, it is necessary to test for frequencies of rotating fields that vary with distance. Such data is probably not available, but does not appear likely that such a condition is possible, since frequencies of radiation do not vary with distance.

For Concept #2, the radial velocity of the rotational field exceeds the speed of light beyond a certain distance from the center of the hydrogen atom. Such a result seems unreasonable by virtue of Einstein's theory regarding the limited speed of light. The same applies to Concept #3 although, this concept could serve to provide a degree of insight as to the mechanism that produces the speed of propagation of light. Thus all of the possibilities under consideration are enigmatic.

How are these issues to be resolved? Is it really possible for a rotational wave to exceed the speed of light? If it doesn't then we have a new dilemma. In order to ensure that the velocity of the field never exceeds the speed of light in any direction, the rotating dipole must have a distorted rotating field which winds up on itself like a nonlinear spring. The concept of a stationary field which changes shape with time may be harder to accept than the possibility that the wavefront rotates faster than the speed of light.

If the speed of light were to be exceeded in this manner, would it violate any of the basic laws of physics? Einstein's theory applies to the radiation of energy, which is the transmission of energy through space. The rotating field is a constant field and does not radiate energy. It is in the form of a dynamic spatial energy distribution with fixed characteristics. Maxwell's equations for electromagnetic radiation have been examined by many scientists, and they show that energy is radiated only under certain specific conditions. For instance, an antenna radiates energy, while an electromagnet does not propagate energy. Therefore, it does not appear that either Concept #2 or #3 violates any of the basic laws of physics, except the Theory of Relativity.

Bohr described the conditions under which energy transfer occurs for the hydrogen atom, wherein stationary states (the rotating dipole examined above fits in this category) do not radiate energy, while a change in the orbital radius produces radiation. Therefore, if we accept the possibility that a rotating field might _not_ radiate energy under certain conditions, then it is also possible that a rotational wavefront can move faster than the speed of light (in the direction of rotation). This assertion is testable by conducting on a rotating dipole.

But if the rotating dipole does not radiate energy, then how is it possible to measure the rotating field? Well, we can measure the characteristics of a magnet, and it does not radiate energy, and a dipole has characteristics similar to a magnet. Therefore, it should be possible to measure the field, even if it does not radiate energy.

If none of these concepts hold up, then we would have to conclude that Rutherford's model has a major flaw. While this model is known to have some problem areas of low consequence, it is still the best and most accepted model available. Altering Rutherford's model in order to relieve the dilemmas is not a trivial problem, and the model would require extensive changes in order to accommodate the limited radial velocity of the field (back to "plum pudding?"). On the other hand, it does not appear that much research has ever been done on rotating dipoles, especially in measuring the angular (rotational) speeds of the rays, so the door is open.

Concept #3 is interesting in that it provides a possible clue to the mechanism that produces the delay in energy transmission. But then it also infers the presence of something that bends the electrical field as it rotates. The bending of the field requires a cause, which would lend substance to the *ether* theory. In any case, we can expect that the universe contains the aggregate of a enormous number of moving electrical spatial fields, each of which is low in level at far distances.

We have one other piece of evidence from which to form conclusions. It is known that our space environment contains a noisy electrical field. The radiation of space has been measured many times and low level noise is known to exist everywhere. One possibility is that the ether, if it exists, may be an aggregate of noisy electrical fields of this type. Light is delayed in passage through an optical window, such as a window pane, and the amount of delay is proportional to the thickness of the window. The delay accounts for the reduction in the speed of light while it is in the window. Photons must therefore be interacting with the fields of the rotating dipoles of the atoms that make up the window. Perhaps the remote electrical fields of the mass of the universe are exerting a similar, but lesser, effect, thus accounting for the delay in transmission of light through space. Einstein proposed that a beam of light would be slowed and deflected by the gravitational field of the mass of a planet near which it passed, and his theory is now accepted due to earlier measurements which confirm the theory. It is possible that the noisy electrical field of space is the mechanism that produces the delay in transmission of electrical radiation?.

Our investigation of the hydrogen atom, as a rotating dipole, has led to some important contradictions, enigmas and dilemmas. The concepts listed above appear to be testable, to some degree, but we do not have the resources available to pursue such an endeavor. To go further, we must choose one of the above concepts as the most likely possibility. Concept #1 has too many difficulties associated with it and will be abandoned. Concept #3 is similar to Concept #2, but the field of Concept #3 has an additional distortion which is not known, for sure, to exist. The effects of field distortions can be considered after the dilemmas and enigmas of the rotating dipole are resolved. Therefore,

Ch. VIII - 110

we will choose Concept #2 as the basis for determining the nature of the external force field of the atom. We can now proceed with the full revelation of the secret of gravity.

And it must follow, as the night the day.

---Shakespeare

CHAPTER IX

The Gravity Equations

The manner in which the gravitational force is produced is described in this chapter. Further evidence is necessary to substantiate the assertions of the previous chapters. The method of deriving the gravitational force solution is necessarily mathematical; however, one of the goals is to present information as simply and clearly as possible, and the mathematical solutions are given in the Appendix. The difficult task of describing the solutions to the complex set of equations representing atoms in three dimensional space, in simple terms, will now be attempted.

As before, the hydrogen atom has been selected for this example since it is the most simple atom with only one electron and one proton. Even with this simplification, the problem is complicated by the lack of an exact model of the atom itself. The orbital path of the electron is directly related to the gravitational force, and the precise shape of the path of the electron is not known at this time. Potential electron orbits are presented in Chapter XVIII, but it would not be practical to assess all of the possible orbits, and Rutherford's orbit is selected for convenience. The orbits are planar, and the basic combination of orthogonal orbits of two atoms (in three dimensions) will be considered.

The basic approach to solving the gravity problem was given in Chapter VII. The electrical force within the atom is much stronger than the gravitational force (10^{39} times as strong), and we must show how the weak gravitational force, external to the atom, is produced by the electrical force. It is currently believed that the electrical atomic forces cancel external to the atom, but even a very slight nonlinear condition can produce an incomplete charge balance, and the assertion is that the external field of the charges within the atom do not completely cancel, external to the atom, thereby producing the gravitational force.

But, even with this straightforward approach, another serious problem exists. Neither electrical nor electromagnetic fields exhibit the isosymmetric force properties of gravity; each is symmetric in only one plane. The process by which isosymmetry is exhibited will also be examined.

The force of attraction between two *hydrogen* atoms will be used to develop these equations. The gravity equations, presented in the appendix, describe the attractive forces between two separated atoms, and the attractive force approaches Newton's gravitational force equation at distances which are much greater than atomic distances. The electrical forces within the atom act in a similar manner, and this correlation is fundamental to the concept of a universal field theory.

The various analogies between electrical and mechanical systems is well known. In the field of mechanics, the analogous mechanical model for the two hydrogen atoms is a dynamic four-body mechanical system with coupling between all four bodies. In the field of electronics, a simplified two-dimensional system correlates to two coupled oscillators. The analysis of coupled oscillators provides further insight as to the characteristics of atoms. Since the atom is a three-dimensional spatial system, the analysis is more complex due to the coupling of forces between all of the components of the atoms. All of forces are summed for each orientation of the atoms.

Before beginning the analysis, the current rationale as to why gravity is not electrical is examined. (See Halliday and Resnick's text *Physics* listed in the bibliography. Their book was chosen for the reference, since it is the clearest presentation that was found). They analyzed the electrical forces of the hydrogen atom and compared them with the gravitational forces, thereby obtaining a ratio of the two forces. The conclusion that results from their analysis, is that *gravitation cannot be electrical in nature*, which is the belief that is currently accepted by the scientific community. This conclusion is based on two important factors:

1. In every atom, there is an electron for every proton, and

the neutron is accepted as being electrically neutral..

2. The attraction of the electron to the proton is calculated to be about 10^{39} times as high as the gravitational force of attraction of their masses.

By virtue of (1.) above, there would be no attraction between hydrogen atoms due to the cancellation of electrical forces external to the atom. Because of (2.), the difference between the electrical force and the gravitational force is so great that the disagreement in levels cannot be resolved by any relationship between these forces. The isosymmetry problem, mentioned in more recent texts on the subject, is also an important consideration. No known electrical phenomenon exhibits isosymmetry, which further substantiates their conclusion. Although various schemes describing possible ways in which isosymmetry could occur have been proposed in the past, the source of the gravitational force has usually been attributed to sub-atomic particles, and no one has yet proffered any theory that is commonly accepted.

Other than these two serious difficulties, the similarities between the electrical force and the gravitational force are notable. In the Rutherford experiment that was described earlier, alpha particles from a radioactive source were directed at a gold foil of having a thickness of about 0.1 micron. Using the measurements of scattering angles, he derived an equation which fits the results of his tests. In the Rutherford model, the repulsion forces <u>inside the atom</u> are inversely proportional to the square of the distance and are due to a Coulomb (electrical) force field. The gravitational force is also inverse square law. In other words, the internal atomic forces vary in a manner similar to that of the gravitational force except that they are much higher and are electrical in nature.

Nevertheless, scientists have dispensed with the notion that the gravitational force can be created from the electrical charges within the atom because of the problems mentioned above. Following the reasoning of Einstein, gravity is still considered to be a separate phenomena, and the Newton formula for gravitational force between mass is the accepted current model for gravity. In order to dispense with the notion that gravity is a phenomenon by itself, it will be necessary to resolve these major difficulties and to provide solid evidence that the gravitational force can be produced by the electrical forces within the atom.

With such an enormous difference in the degree of these forces, it seems surprising that highly accomplished scientists would not look for a secondary phenomena. Any nonlinearity can produce error terms for analysis in the space of mathematical functions. Even a <u>very slight</u> variation in the linearity of electrical force versus distance relationship allows the possibility of a correlation between the two forces (gravitational and electrical). The necessary nonlinearity (of the electrical force equation) is present and is well known. The mathematical formula for the force between electrical charges is described by Coulomb's Law,

$$F_e = q_1 q_2 / 4 \pi \varepsilon_0 r^2 , \qquad \text{[IX-1]}$$

which is a nonlinear equation (due to the $1/r^2$ term). Thus all of the necessary ingredients to produce the gravitational force from the electrical force are present. Most of the forces in nature that are not gravitational are electrical. It is possible that <u>all</u> non-gravitational forces are electrical, but some are classified according to other characteristics, such as radioactive forces. Therefore, once the gravitational force is identified as a consequence of the strong electrical forces within the atom, all forces reduce to a common force, as was contended in Chapter VII.

The gravitational force of Newton's equation is calculated in Appendix I. The gravitational force varies with the distance between two hydrogen atoms, as pictured in Figures IX-1 and IX-2. Note the force between the two hydrogen atoms at a separation of 0.1 meter. This computation is used for comparison with later calculations.

$$F_{gravity} = 1.8687 \times 10^{-62} \text{ newton} \qquad \text{[IX-2]}$$
$$= \text{gravitational force between two hydrogen atoms } 0.1 \text{ meter apart.}$$

The net <u>electrical force</u> between the two atoms has been calculated for this same distance in Appendix II. The Rutherford model of the atom, which pictures the electron rotating about the proton of the

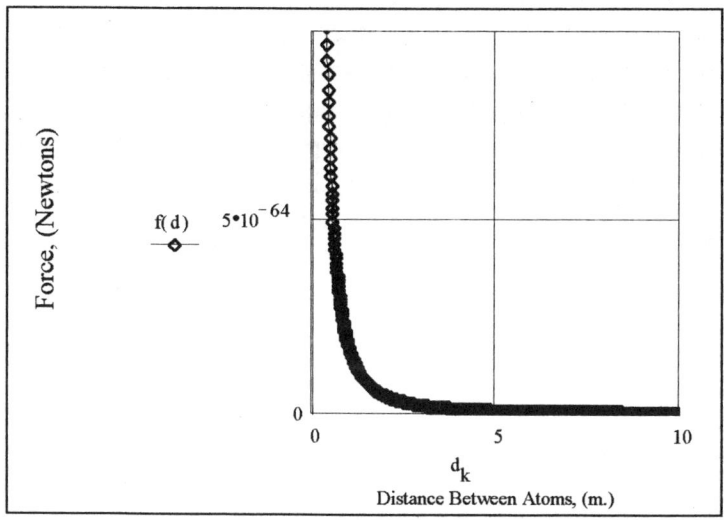

Figure IX-1. The Gravitational Attraction Between Two Hydrogen Atoms (Linear Scale)

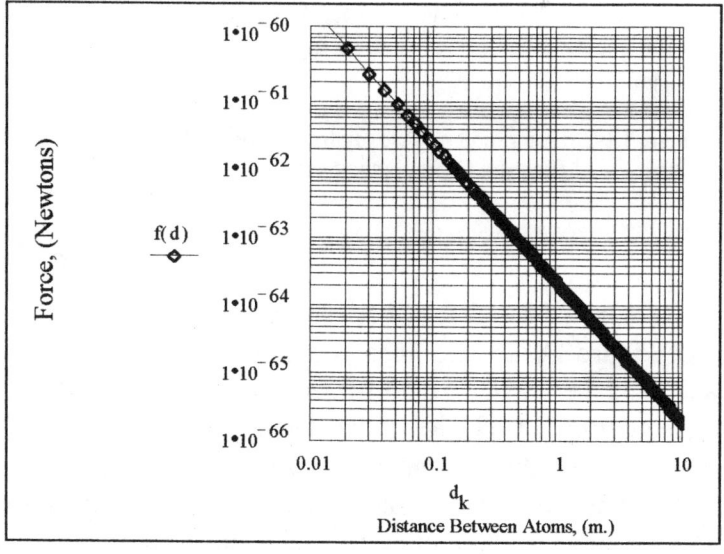

Figure IX-2. The Gravitational Attraction Between Two Hydrogen Atoms (Logarithmic Scale)

hydrogen atom, will be used for the analysis. The planar orientation of two hydrogen atoms, illustrated in Figure IX-3, shows two hydrogen atoms that are separated from each other. There are four forces acting on the electrons and protons of these atoms (excluding the internal forces of the atom). A repulsive force is present between like charges, but an attractive force exists between opposite charges. This electrical system is analogous to four-body mechanical system with nonlinear springs to simulate the nonlinear force between the electrical particles.

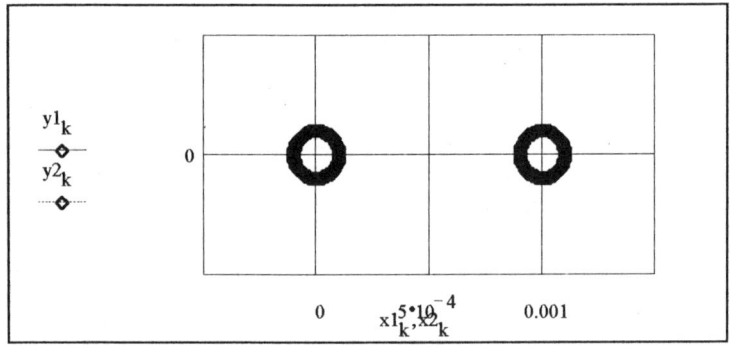

Figure IX-3 Illustration of Two Hydrogen Atoms With Electron Orbits in the Same Plane

If the sum of the forces between each of the four electrical particles of Figure IX-3 were linear with distance, then the charges would cancel and there would be no attractive force between the atoms. However, equation [IX-1], describing Coulomb's law, provides the nonlinearity that provides incomplete cancellation of forces.

For example, consider two atoms whose nuclei are separated by only a distance equal to three times the radius of the electron orbit. If the electron approaches the near vicinity of the proton of other atom, the attractive force between them becomes very strong. When the

Ch. IX - 117

electron rotates to a position on the opposite of the atom, the force is only about one tenth as strong as it is on the strong side. The average force of attraction of the electron to the external proton is more than twice as strong as the repulsive force of the two protons, and the unbalance of force is very pronounced. In this example, the distances chosen approach those of atomic and molecular lattice spacings, and as the distance between atoms increases significantly, these attractive forces decrease to a very small value. Therefore, for greater separations at which gravitational forces are measured, the gravitational forces are produced by the very tiny differences in the sum of the electrical forces between the atoms as averaged over a rotation of the electron about the proton.

The sum of the all of the repulsive forces between like charges must be subtracted from the sum of the attractive forces in order to determine the overall gravitational force. For close separations the force is great, while at large distances the force is very small. If it can be shown that the resulting average attractive force is sufficiently large to provide the strength of the gravitational force, then the evidence will be strongly supportive of this theory. The equations included in the appendix are offered as proof that this theory is quite believable.

Unfortunately, space has three dimensions, and since the chosen orbit is planar, the various orientations of the two atoms must be considered. The orbit of either of the two hydrogen atoms can exist in any one of three orthogonal (at right angles) geometric planes or any intermediate angle. Another variable is the relative rotational positions of the electrons in their respective orbits, which is another complication. Consequently, several physical states are analyzed. Since the exact orbital pattern of the atom is not known, the approach is to study the three orthogonal arrangements of the two atoms and then deduce the results that are obtained by the intermediate orbital positions.

The instantaneous force between the two separated atoms is the sum of the four forces between the individual charges within the atom. The four forces do not cancel as they rotate, and it was found necessary to scale the radius of the actual electron orbit in order to minimize computer simulation errors. For electron orbits whose planes are aligned, the overall scaled instantaneous force is pictured in Figure IX-4. Two cycles of orbital pattern are illustrated to show the cyclic repetition. The average force is obtained by summing all of the forces

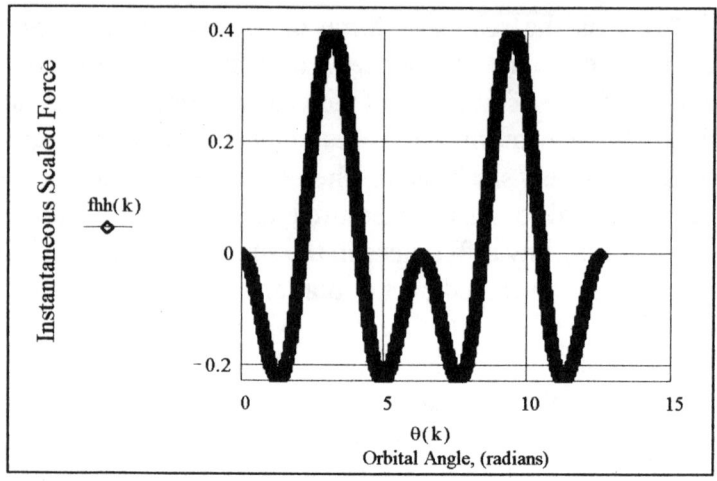

Figure IX-4 Instantaneous Force Between Two Hydrogen Atoms
With Their Orbits in the Same Plane

as the electrons rotate over the two cycles. The result is shown in Figure IX-5.

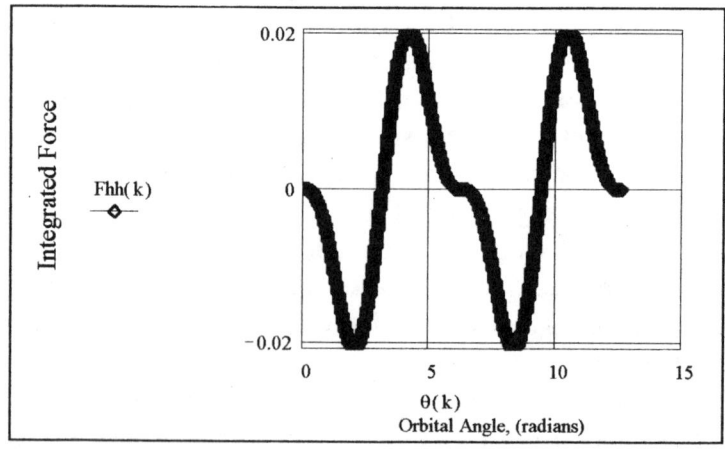

Figure IX-5 Integrated Sum of the Four Forces Acting Between Two
Hydrogen Atoms With Their Orbits in the Same Plane

A very small net force of attraction exists, although it is too small to be seen on the graph. <u>Not only does a force of attraction exist, it is much larger than the gravitational force</u> of equation [IX-2],

$$F_{planar} = 1.9348754 \times 10^{-47} \text{ newtons.} \quad [\text{IX-3}]$$

Orbits which are <u>axially aligned</u> as in Figure IX-6 are next considered.

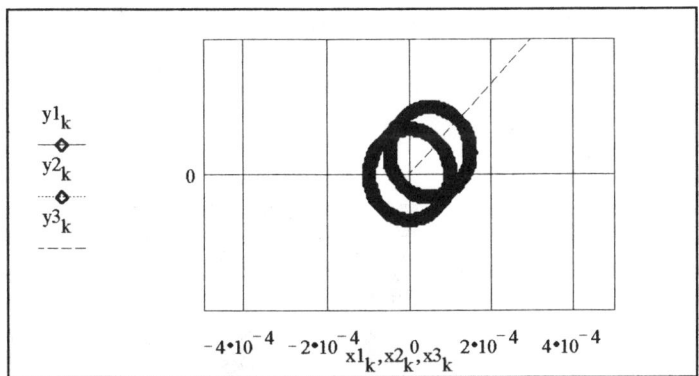

Figure IX-6 Illustration of the Configuration of Two Hydrogen Atoms With Planar Electron Orbits Aligned Axially

Again, two cycles of repetition are analyzed, and the resulting overall instantaneous force is illustrated in Figure IX-7.

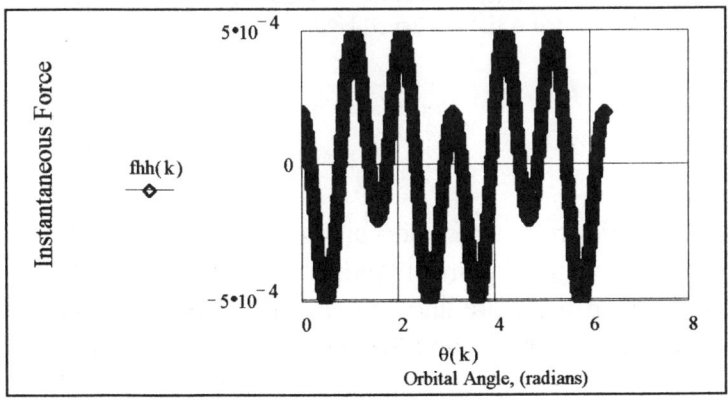

Figure IX-7 Instantaneous Force Between Two Axially Aligned Hydrogen Atoms

As before, the force levels repeat for each of the two cycles, while the average force does not sum to zero at the end of the second cycle. The curved of the integrated (summed with rotation) force is shown in Figure IX-8.

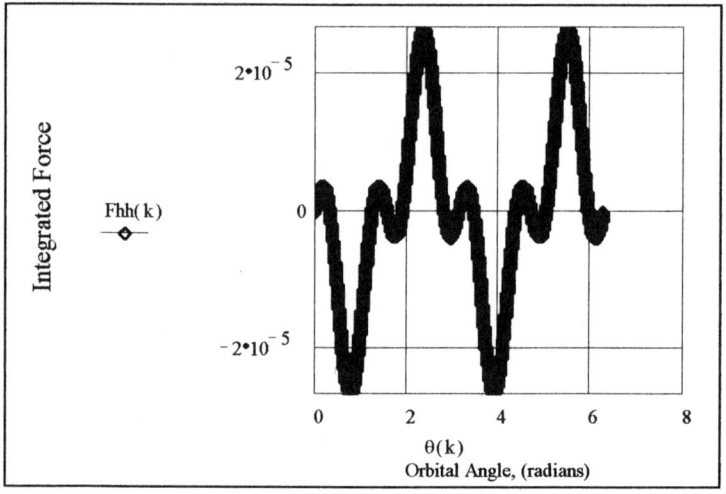

Figure IX-8 Integrated Sum of the Four Forces Acting Between Two Axially-Aligned Hydrogen Atoms

For this geometric arrangement of orbits, the average force is much less than that of the planar alignment, and, surprisingly, it is *negative*. The two atoms would therefore line up with their planes coincident rather than with a common axis if no other forces acted on the atoms.

$$F_{axial} = -2.586374 \times 10^{-50} \text{ newton} \qquad [IX-5]$$

Ch. IX - 121

Orthogonal orbits for the hydrogen atoms are pictured in Figure IX-9.

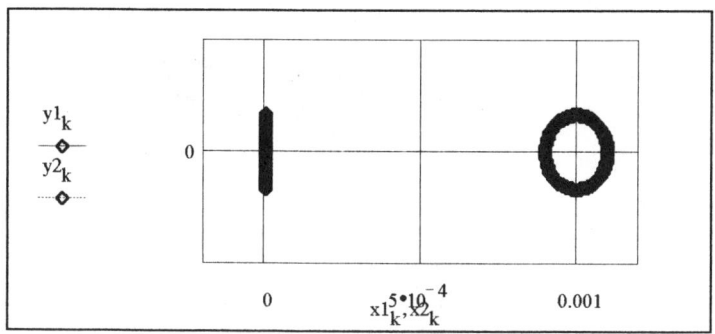

Figure IX-9. Ilustration of the Configuration of Two Hydrogen Atoms With Their Orbits Aligned Orthogonally.

The shape of the waveform of the instantaneous force of Figure IX-10 is somewhat different from that of the two prior geometric configurations.

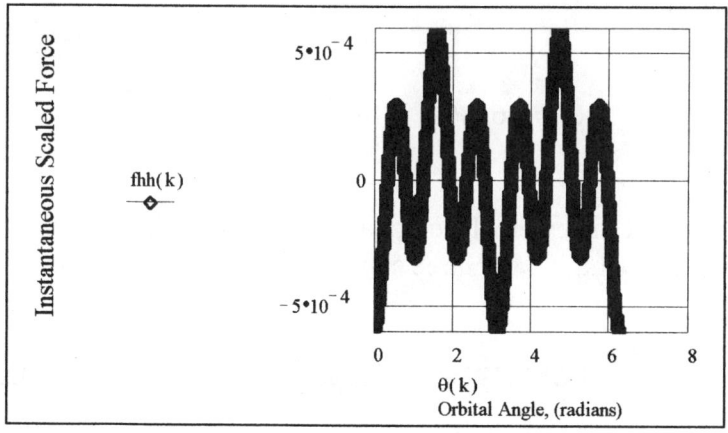

Figure IX-10 Instantaneous Force Between Two Hydrogen Atoms WithTheir Orbits Aligned Orthogonally

The summed force is pictured in Figure IX-11.

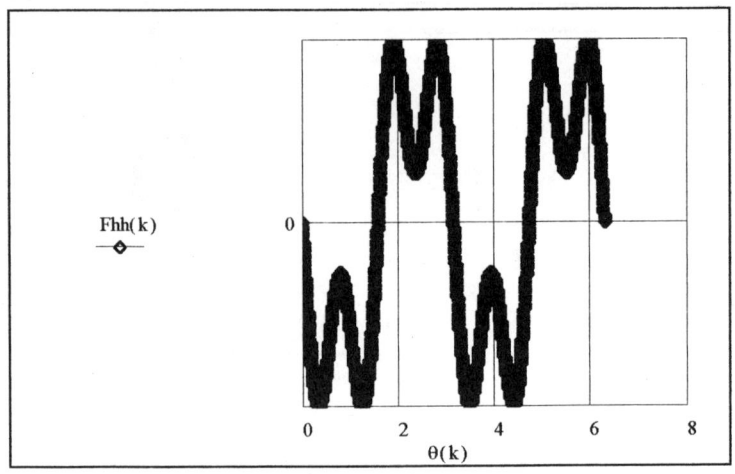

Figure IX-11 Integrated Sum of the Four Forces Between Two
Orthogonally-Aligned Hydrogen Atoms

The average force is attractive, similar to that of the planar alignment, although somewhat lower in strength.

$$F_{orthogonal} = 1.2906286 \times 10^{-47} \text{ newton.} \quad [\text{IX-6}]$$

An estimate of the external electrical force between hydrogen atoms is obtained by averaging the three summed forces.

$$F_{electrical} = 1.0743392 \times 10^{-47} \text{ newton.} \quad [\text{IX-7}]$$

This is not a highly accurate estimate, but it is sufficient for our purposes.

Comparing equation [IX-7] with equation [IX-2], the average electrical force between atoms is seen to be considerably higher than that of the gravitational force. <u>The existence of a significant electrical force disproves the conclusion that the gravitational force cannot be obtained from the electrical forces within the atom.</u>

However, the gravitational forces on a mass are isosymmetric in all three planes through which the body is rotated. The flat orbit that was used in the above example is symmetric in two planes but not in the third. Is it possible that mass is not always completely isosymmetrical? Isosymmetry is only necessary for one of two masses in order to have the force of attraction be independent of the spatial orientation of either of the masses. The earth is believed to consist mainly of iron, and the iron atom may be more symmetric in all three axes than other atoms. Cavendish's experiment, you will recall, used iron balls to verify Newton's equations of gravitational force.

We know that some materials tend to repel others, while different materials may have an unusually high attraction for one another. This phenomena has been attributed to the forces of static electricity. It is also possible that the lack of total isosymmetry of certain atoms may also produce a similar effect.

Another possibility is that the electron of the atom moves in a path along a sphere, in which case the force would be isosymmetrical. Again, the shape of the electron orbit is an important factor. We will investigate electron orbits in Chapter XIV.

The electrical force of attraction between two hydrogen atoms was shown to be much higher than the gravitational force (for the Rutherford model of the atom). Even with a globular or spherical orbit, the electrical force is still much higher than the gravitational force. Two possibilities for resolving the force difference come to mind. Either the gravitational force is stronger than previously thought, or the model of the atom must be altered. If Newton's equation is exact, then the low gravitational force must be produced by an orbit that differs in some way from the Rutherford model. Few scientists have questioned Newton's Theory of Gravity. Some of the problems with gravity will be discussed in Chapter XII, but first let us see if the Rutherford model can be modified such that the electrical force equals the gravitational force.

One method of altering the model of the atom is to allow the proton to rotate. If the proton rotates in a circle with a radius the same as that of the electron, the external electrical forces should cancel. In that case, the electron and proton would rotating about the center of the dipole. A slight rotation of the proton is already believed to exist (purportedly due to inertial forces). By properly adjusting the radius of movement of the proton within the hydrogen atom by a significant amount, the electrical force will equal the gravitational force. The mathematical analysis of the external electrical forces generated by a hydrogen atom, for a modified model of the atom wherein the proton rotates, is presented in Appendix III. With a path radius for the proton which is about 70% of the electron radius, the external electrical force between atoms reverses and becomes repulsive. This figure is not exact, however, due to the limited accuracy of the numerical computation method. The gravitational force is so low that it falls within the "sigma" noise errors of the program. Therefore, it was not possible to determine the exact radius of the proton orbit to make the external electrical force exactly equal to the gravitational force, but a more powerful computer with a simulation program of greater accuracy will produce the proper result.

The above classical analysis of the external forces exerted upon atoms, based on the electrical charges within them, has produced additional information about the dynamics of atoms that is difficult to envision by using the energy concepts of quantum physics. The importance of the determination of the orbital pattern of electron movement within the atom is also placed at a position of greater significance because of the above analysis.

But what about the neutron? A neutron consists of a proton and an electron, and it is small enough that it can reside in the nucleus of the atom. A model of the neutron can be constructed rather neatly on the basis of the arguments presented above. The rotation of the proton allows the neutron to fit will within the diameter of its circle of rotation. If a hydrogen atom is squeezed down to a smaller size such that it fits properly within the nucleus of the atom, the electrical force of attraction is reduced. Thus the neutron can be modeled in a manner similar to the way the hydrogen atom was modeled. Since we have plenty of electrical force to work with, the radius of the path of the proton within the neutron can also be reduced, thereby offsetting the excessive reduction in electrical force that occurs when the orbital radius of the electron. Therefore, a neutron, which is much smaller than the size of the atom, can still produce the required gravitational force in spite of a much smaller orbital radius.

This hypothesis also tends to support the idea of black holes in space. The *available* electrical attractive force of a hydrogen atom is higher than the gravitational force. According to the equations, there is a limit to the amount of condensing that is possible before the attractive force decreases. Nevertheless, squeezing hydrogen atoms into neutrons permits many neutrons to be crowded together, forming a mass of high density. Observations indicate that neutrons generally do not like to congregate (neutrons, by themselves, tend to be unstable), but special conditions could conceivably exist in outer space that permit them to condense into a huge mass. This argument does not verify the presence of black holes in the universe; only that the hypothesis does not contradict the possibility that black holes might exist. It does not support the idea of a singularity within the black hole, however. The forces within the black hole have a limit since the number of atoms, and the space that they fill, is finite.

Some new questions come to mind when considering the electrical gravitation theory:

1. What is the shape of the orbit that produces the gravitational force?

2. What is the nature of the force of electrical attraction between the electron and the proton, holding the electron in its orbit?

We will try to get some clues as to (1.) in the following chapters. The second question is much tougher. Perhaps the characteristics of moving fields (waves) have particular characteristics that only allow certain conditions to exist. We do not have enough information, as yet, to be able to form reasonable conclusions and have reached the limit of knowledge of inner space.

The fundamental argument of the above hypothesis is that the external electrical forces of atoms are more than sufficient to account for the gravitational force, and that it can be made exactly equal to the gravitational force by adjusting the radius of the orbit of the proton as it rotates within the nucleus of the atom. The secret of the source of the gravitational force has thus been revealed and verified to a reasonable degree of accuracy.

It should also be mentioned that an attempt to solve the complex equations of Appendix II directly resulted in a solution which contains the square of the radius of the atom in the numerator. A direct solution to the electrical equation shows that mass is proportional to r_e^2, the square of the electron orbital radius of the atom. It has been argued that the orbital path is an important factor in determining the external electrical forces of an atom. Choosing another orbital path can produce a somewhat different result. It may be possible that a spherical orbital path produces a force of attraction for which *mass is proportional to r_e^3, the volume of the atom*. Such a result would fit in quite well with other theories of chemistry and physics. Atomic orbitals will be explored in Chapter XIII, and the radiation from changes in orbital paths will be examined in ChapterXIV.

Assertion, unsupported by fact, is nugatory.

---Junius

Ch. X - 127

CHAPTER X
Electrical Conduction and Magnetism

Electrical Conduction:
Electricity is of importance in understanding the universe, and we should therefore understand it thoroughly. Although the principles of electricity have been successfully applied to numerous practical applications in various disciplines, much of the theory still remains mysterious. For instance, various theories regarding the conduction of electricity have been developed over the years, and all of them have various flaws. A more well-defined theory will be presented, but first let us examine some of these theories.

Free Electrons:
This theory is at least 50 years old and attributes conduction to "free electrons" that are present in a conductor. If there are free electrons in a material then where are they located? They must be evenly distributed and free to move, since an uneven distribution produces a field, and a field produces current in a conductor. It is not likely that they would be attached to any atom as *extra* electrons, since there must be an electrical force holding it to the atom (by attraction to a proton) to achieve a balance in force. If there is an unbalance in force, then a preponderance to lose (or gain) electrons would be expected, and the material would act like a semiconductor.

Electron Cloud:
This more recent theory pictures an "electron cloud" around an atom from which electrons can be drawn for conduction. But even the little hydrogen atom, with its single electron, is believed to have a cloud of electrons surrounding the nucleus. It is obviously impossible for a single electron to form a cloud, and therefore the picture of an electron cloud is just be an illusion created by a measurement problem. Quantum theory is based on statistics, and we cannot perform experiments on a single atom in a fast enough time to capture the picture of a single

orbit. Therefore we must be seeing the effects of many orbits in many atoms in the measurement (the results are "clouded").

Overlapping Energy Bands:

Quantum theory defines conductors and insulators in terms of atomic energy bands. The electrons in an atom are revolving about the nucleus in orbits of differing energy levels. If these energy levels are separated, then the material is an insulator. In order to have conduction, the energy bands must therefore overlap since a voltage of any level can produce conduction, which would not be the case for energy gaps. This explanation implies that the electrons are free to move from one orbit to another (within an atom) since the energy level is related to the radius of the electron orbit. Therefore, a continuous energy band would produce a radiation spectrum is continuous. A continuous spectrum results for "blackbody radiation," but not all blackbody radiators are good electrical conductors. Energy bands reflect the nature of the atomic orbital electron motion, which does not <u>necessarily</u> infer conduction. If two atoms are directly adjacent to one another and then pulled apart, then their electron energies would not vary significantly, but their ability to interchange electrons (conduct current) must change drastically.

Electron Gas:

This version is more realistic. The electrons in a conductor are thought to be in motion, moving from atom-to-atom in a random manner and thus producing the illusion of a gas (gases exhibit this type of motion). The application of an electrical field therefore orients the flow and produces conduction. This scenario also fits the energy band overlap concept, but with the important additional requirement of interchange of electrons between atoms. This theory does not, however, explain why the electrons are moving from atom to atom. If it is heat that produces the motion, then conduction should stop at a temperature of absolute zero, and this is far from the case.

Ch. X - 129

These conflicts illustrate the confusion that results from having a group of theories, none of which fully correlate with physical phenomena. For a more rational explanation of electrical conduction, the electromechanical description of atoms, as presented in Chapter V, and the Periodic Table of the Elements of chemistry can be utilized to locate conductors. The following assertions apply to electrical conduction:

1. Atoms which have a high degree of *dynamic mechanical balance* do not conduct electricity well. If the atom is highly balanced, the material tends to form a gas since the molecules have no inclination to unite with one another. The noble elements, such as neon, radon, etc., have this property and are not conductors. As the balance of an element decreases, the molecules tend to unite, forming liquids and then solids. Since conductors are solids, conduction requires some degree of dynamic unbalance (ionic gases and liquids can also conduct, but by another mechanism). This places conductors towards the left side of the Periodic Table of the Elements.

2. Too much *unbalance* tends to reduce conduction. Hydrogen is the most unbalanced atom, and it tends to zoom through space at supersonic speeds, bouncing around and crashing into other atoms that it encounters. It is not in one place, next to others atoms, long enough to produce conduction. Therefore, conductors will be found away from the both of the end columns of the Periodic Table of the Elements.

3. The chemically active elements are not very well-balanced and can be eliminated. This places the conductors further toward the center of the table.

4. The remainder of the chart now consists of metals and semiconductors. We know, from experience, that metals will conduct better than semiconductors, and the "packing" of the atoms in the crystal will have an effect.

The atoms must be close enough together to allow the electron to move from one atom to the other. The best conductor will have a degree of dynamic stability, not be highly chemically active, and have close packing. The tightest packing of a crystal is the *cubic closest packing* structure, each atom touching twelve others. In chemistry, the "radius ratio rule" can be used to determine which atoms form a cubic crystal structure (this rule does not always apply). An atom will touch at least eight of its neighbors if the ratio of its positive ion to its negative ion if the ratio is greater than 0.732. The metals *silver, aluminum, calcium, nickel,* and *lead* fall in this category. Calcium is chemically active and can be eliminated. The others we already know to be good conductors.

5. Now that we have located the conductors in their general location in the Periodic Table of the Elements, let us examine the electron configurations of these elements. The electrons for each element exist in energy shells about the nucleus. Each shell can hold only a maximum number of electrons. *The best conductors have one portion of the electron configuration in common.* As the shells fill up with electrons they form *the equivalent of a noble gas, followed by an unfilled spherical shell*. The conductors can easily be selected from the table by simply searching for the spherical s^1 shell followed by a filled shell (the noble atom structure). The unfilled spherical shell is missing exactly one electron and one proton, which unbalances the atom slightly. The other shells give the atom a sufficient degree of balance that it can come into contact with adjacent atoms and thus easily transfer electrons.

The good conductors are metals and can be determined by following the above procedure. The above theory provides rational reasoning as to which atomic features are necessary for conduction. It also explains another phenomena: the conductivity of a good conductor decreases with temperature. Atoms and molecules become agitated with temperature. This vibrational effect has already been expressed in the form of an equation which has proven to be quite accurate. The above arguments provide a reason why the resistance of a good conductor increases with temperature while insulators tend to conduct at the higher temperatures. A slight amount of mechanical unbalance is required for conduction. There are exceptions to this rule, and, for those cases, the atoms must be so tightly bound that a higher temperature is required to make them vibrate slightly.

A conductor can also be described as a material that is just as likely to give up an electron as it is to accept one. When atoms in close contact begin to vibrate, it appears inevitable that an exchange of electrons will occur. Such an electron movement does, in fact, occur under certain conditions. When a conductor is heated to an elevated temperature, a phenomena called *thermionic emission* occurs, wherein the electrons achieve enough energy to escape from the surface of the metal. They do not actually escape very far into the atmosphere but are held in the vicinity of the surface. Richardson, in England, developed an equation for the rate of electrons (electrons/second) escaping from the heated surface. Extrapolation of this equation suggests that *superconduction* (having exceptionally low electrical resistance) would be produced near absolute zero for certain conductors. Superconduction is known to occur at low temperatures, but only for certain metals (and certain compounds). Superconductors are known to have odd properties. Aluminum, lead, zinc, mercury, tin, etc. become superconductive at a few degrees above absolute zero.

Some other metals become superconductors at much higher temperatures. These turn out to be elements which have 3, 5, or 7 valence electrons outside of filled shells (instead of a single electron for the conductors). Note that these are, again, unbalanced atoms (compared to an even number of electrons in the unfilled outer shell).

The *semiconductors* fall in the region of the Periodic Table that conductors are located but a specific semiconductor will either accept electrons readily or give them more easily. Thus a semiconductor material is classified as either an *acceptor* or a *donor*. Since neither type accepts electrons to the same degree that it gives them up, they are not good conductors. However, by combining both types of semiconductors into a common material the conduction can be controlled. Transistors are fabricated by using separating layers of the two types of semiconductor material, and adding terminals (to the semiconductor chip) for the conduction of current.

The conduction of current in semiconductors can also be explained in terms of moving electrons and "holes," the hole being a spot in an empty spot in an atom or molecule where an electron would normally be located. A similar process must also occur in conductors. If two atoms nudge each other with enough force, the electrons can push one another with enough force to eject either or both of them from orbit. In order for conduction to take place, the hole, or missing electron, must be re-filled. If a voltage is impressed across a conductor it pulls one of these dislocated electrons from one end, while at the same time trying to force an electron out of orbit at the other end. This process correlates with the "energy band overlap" of quantum theory.

Any theory of conduction must also be in accordance with Maxwell's equations. If cause and effect can be separated, then, using Maxwell's equations as a basis, perhaps the picture of the field will become more clear. A simple explanation of these equations will now be presented:

Conduction evidently begins with only one or two electrons; that is, it is never instantaneous. The limited buildup of electrical current is controlled by the conditions of the path through which the electrons and holes move. The electrical parameter that determines the rate of current increase for a given voltage is *inductance*. In order to understand conduction and magnetism we need to have an idea of what we mean by inductance. The following is a very simplistic analogy:

> Imagine a huge racetrack with many cars which simulate electrons. The cars are all moving, mostly in small circles, but with no common direction. As the race begins, a gate opens and only the cars in front can move. The few that happen to be moving in the right direction go first, followed by others in the front of the pack. At the back of the pack another gate is simultaneously opened allowing more cars to enter, and the cars from the front are eventually move in a circle eventually

reaching the back of the pack. The average speed of the cars around the track builds slowly but continuously, but some cars are always moving around in small circles while the pack is beginning to move around the track. When the race is flagged as being over, the cars cannot stop immediately and slow down gradually. The longer the track, the slower the buildup of speed of the cars. The track is analogous to the inductance of an electrical circuit. The longer the track, the longer it takes for the speed of the cars to build, and the longer a conductive path, the higher the inductance and the longer it takes for the current to build.

The above analogy is not exact because atoms are indestructible, and low energy collisions do not cause destruction. Also, the electrons move from atom to atom by virtue of the external forces exerted upon them rather than being steered internally. In the electrical circuit, voltage is the prime mover, and it pushes at one end and pulls at the other. When a circuit is circular in form, with electrons entering one end and exiting at the other, what is the cause and what is the effect? Do the electrons start everything moving by virtue of their excess energy pushing them across the first few nanometers of distance, or are they more likely to be pushed or pulled from the other end? Presumably the electron exits first since it has the vibrational energy that is produced by temperature. The *cause* must begin before the *effect* occurs, and this condition probably has never been measured.

By this analogy, it takes time for the number of cars to begin moving in the proper direction, and inductance produces a similar lag in the buildup of an electrical current. The inductance of a circuit depends on a number of factors, the most important of which is *length*. A little arithmetic will be employed at this point. In the most simple case of a long, straight wire, the inductance is

$$L = \mu_0 \, l / 8 \pi \quad \text{(l is the length of the conductive path)}, \quad [X\text{-}1]$$

and in free space the magnetic permeability constant is

$$\mu_0 = 4 \pi \cdot 10^{-7} \text{ H/m} \qquad [X\text{-}2]$$

Ch. X - 134

(H is the unit of inductance in henries).

Therefore, the inductance of a straight wire is proportional to length, so inductance represents the dimension of the length of path over which current flows. The longer the wire, the longer time it takes to build up a current for a given voltage. The permeability constant represents the ability to form magnetic fields. The higher the value of μ_0, the greater the magnetic field. The above value of the permeability constant, μ_0, is a minimum when no other materials are placed nearby and is called the *permeability of free space*. It also represents the *internal inductance* of [X-1] since the wire is isolated in space with no nearby objects. This does not mean that the magnetic field is completely internal since there is also an external field.

What now follows is a unique interpretation of electron flow through an inductor. Again, a little mathematics is again required for this exercise. If the inductor has a negligible resistance (a good conductor), then the equation that describes current flow, produced by a voltage [e(t)], is described by a differential equation

$$e(t) = - L \cdot di/dt \qquad [X-3]$$
$$= \text{inductance times the rate-of-change of current.}$$

Current consists of a fixed number of moving electrons, so this equation can be written in terms of the number of electrons that move within an interval of space,

$$e(t) = - L \cdot \Delta i/\Delta t = - L \cdot \Delta^2 q/\Delta t^2 \qquad [X-4]$$
$$= - \text{ inductance times the rate of change of current.}$$

This equation can now be solved with simple arithmetic. But elementary electrical charges do not change with time. It is the <u>amount</u> of charge that varies. The density of charge varies along the inductor and is a function of distance. Including these parameters, the equation can be rewritten in its most simple form as

$$e(t) = - L \cdot \Delta^2 q/\Delta s^2 \cdot \Delta s^2/\Delta t^2. \qquad [X-5]$$

The "deltas" in this equation represent small amounts of change. The interpretation of this equation is that the acceleration of the electrons and their *location and distribution along the wire* is a function of the applied voltage and the inductance of the circuit. Therefore, <u>inductance describes the amount of inertia of the moving charges</u>. The longer the wire, the greater the inertia. This equation is similar in form to the equation for mechanical acceleration,

$$f = m \cdot a. \qquad [X-6]$$

Voltage and force are the prime movers in these two equations. The *equivalent electrical mass* (of the moving electrons in the wire) is proportional to the length of path times the change in the density of electrons distributed along the wire.

The above analysis implies that electrical charge and mass are closely related to one another, which fortifies the hypothesis of Chapters VII and IX.

Magnetism:

The next question is "What is magnetism?" The concept of magnetism is a scientific invention used to describe an electrical phenomena. The word magnetism stems from the characteristics of the field around a magnet, which is called a *magnetic dipole*. Iron filings in the vicinity of the magnet line up with the *magnetic flux*. But is it really necessary to invent a <u>magnetic</u> field that coincides with the direction of the iron filings near a magnet in order to provide an explanation? If Maxwell's equations are interpreted in another way, then a different spatial geometry can be constructed. Again, some mathematics are necessary to provide the explanation.

Electromagnetic induction is defined by Faraday's law,

$$e(t) = - d\phi/dt \qquad [X-7]$$

where ϕ is the magnetic flux. Voltage is proportional to the rate of change of flux. This equation can be re-written

$$\phi(t) = -\int e(t)dt = L \cdot i(t) = L \cdot dq/dt, \qquad [X\text{-}8]$$

or

$$\phi(t) = L \cdot dq/ds \cdot ds/dt = L \cdot v(t) \cdot dq/ds, \qquad [X\text{-}9]$$

(v(t) is the velocity of the moving charges).

This is the desired form of the equation which will now be interpreted. The variables of space and velocity have been added to equation [X-8], resulting in equation [X-9]. Again, charge does not change with time; but the distribution of charge is changing along the wire. The magnetic flux increases with the length of path of the current (as was true with the previous equation), assuming a fixed charge density within the wire. It also increases with the velocity of the moving charges and the density of charge along the wire.

The term "flux" is artificial. If iron filings are placed in the field of an electromagnet, they lineup, forming a pattern. The filings are said to follow the hypothetical lines of flux in the magnetic field. This is a visual concept, but perhaps not the best description. The above equation shows that flux can be described in terms of the charge distribution along a conductor. For every charge there is an associated electrical field. Therefore, magnetism can be viewed as *another form of the electrical field*, and but a single type of electrical force exists (although it assumes various forms in space).

The analysis of electrical fields by early investigators led to the conclusion that there is energy stored in the field. Electric energy is stored in the electrical field produced by separated charges, while electromagnetic energy is stored in the magnetic field by virtue of the inertia of the moving charges. If a capacitor is charged up by placing a voltage across it, that energy will stay in the capacitor for as long as charge leakage will permit. If an inductor is charged up by placing a voltage across it, it takes time for the electrons to accelerate to a given velocity by virtue of the previously described electromagnetic action. Removing the forcing voltage, however, allows the magnetic field to collapse, and energy is forced back out of the inductor as the voltage is being removed. The dynamic energy storage of a straight-wire inductor is a function of its length, the amount of charge in the wire, and the velocity to which the charges have been accelerated. It is much like the hectic car race previously described, or even like life on the freeway.

Ch. X - 137

Once the cars get moving it is hard to stop them. The same thing happens with electrons.

The above analysis does not conflict with existing theory, but it eliminates the concept of magnetic flux as a separate entity. The hypothetical magnetic flux is not central to the argument and is equivalent to charge distribution and the velocity of moving charges in a wire. Electrical and electromagnetic fields can thus be pictured in terms of voltage, charges, distance, and velocity. Then what about the velocity of an electrical field in a vacuum (the speed of light)? How does that relate to these parameters? It can be stated quite simply:

> Separated unit charges of opposite polarity have a force between them which is given by Coulomb's law (Chapter IX). The electrical force between charges can accelerate the electrons at a rate which is no greater than that which the inductance of free space will allow. Maxwell's equations show that, for the outward radiation of energy, a spherical wave must be generated, and radiation requires a particular physical configuration (an antenna). The limiting factors of electrical-force/charge by the electric constant, ε, and the maximum acceleration of charge by the permeability constant, μ, set a maximum speed for the radiation of energy, and these constants are set by the characteristics of an electrical field in space.

By this analysis, magnetism is viewed as <u>another form of the electric field</u>. The electric field exists between charges distributed along a wire. Since voltages exist along the wire, conduction must also take place. It is therefore a dynamic effect. For a magnet, the dynamics take place within atoms due to the motion of the orbital electrons which is another electrical phenomenon. Each electrical force is thus produced by a <u>single field</u>, an electric field, which establishes the cornerstone of the unified field theory which follows.

The analysis also indicates that the speed of light is limited by the speed of the electrons in the above magnetic flux equation . Therefore, the conclusion is that it is *the speed at which the electrons leave their orbits and move from atom to atom that controls the speed of light*. This speed reaches a maximum in a vacuum where other fields do not interfere.

The radiation of energy can also be produced by a group effect of the electrical charges, which are distributed in accordance with the equations for the inductive and capacitive effects. Radiation of this type is discussed in the next chapter.

The above presentation offers a rational theory for electrical conduction and provides additional information about conductors, while it also creates a vision that fits reality. The theory of magnetism, as another form of the electric field which is dynamic, presents an insight which leads to a description of the fundamental mechanism by which energy is radiated. The information in the following chapters supports these theories without creating conflicts, although some new unexpected enigmas will be shown to exist

What is mind? No matter.
What is matter? Never mind.

---T. H. Key

CHAPTER XI

A Universal Field Theory

Physicists have long sought to establish a unified field theory. The Theory of Relativity was based, to a large degree, on analysis of field problems. Einstein extended the limitation of the speed of light to include <u>everything in the universe</u>. He attempted to formulate a pure field physics and claimed that the field concept was the most important invention since Newton's time. He was unsuccessful in solving the secret of gravity and in establishing a universal field concept, saying "We have two realities; *matter* and *field*."

Mills and Yang, in 1954, published a paper which described a method of accounting for global symmetry of the electrical fields of atoms in six vector fields by means of isotopic fields of individual nucleons (the nucleus of an atom); a rather complex concept, not easily understood.

Then the photon was believed to be the carrier of electromagnetic force, and two new subatomic particles were invented to solve the problems of the various forces. During the decade beginning in 1960, particle physics concentrated on quarks and gluons, and these were thought to be the carriers of the true "strong force." A later effort by Schwinger resulted in a theory wherein the "weak force" and the electromagnetic force were thought to be of the same strength, in symmetry which had been broken, and involved bosons. Most all of these theories, and others that followed, involve subatomic particles in various schemes that combined the forces of matter.

No one has thus far been able to establish a complete theory regarding a universal field, although many have attempted it besides

Einstein. There are substantial reasons to believe that a universal field exists. One factor that supports this belief is that matter is granular, composed of elementary particles, atoms and molecules, and both electric charge and energy are also granular in nature. These basic characteristics lead to the suspicion that the universal field may be of an *electrical* nature.

The similarities between the characteristics of electricity and gravity are well known. For a universal field, either electricity must be a function of gravity, or else gravity is produced by electricity. Einstein equated energy and mass. On this basis, he concluded that energy, in the form of a field wave would be attracted to a mass, so that a light wave would bend when passing near a planet. Measured results reportedly confirm his conclusion. But a light wave is an electromagnetic field, and, using Einstein's logic, mass is therefore also a field.

The relationship between electricity and gravity has always been difficult to resolve. The Theory of Relativity is based on the gravitational effects as being a separate entity, in which case a unified field theory cannot be established without defining the relationship between the electrical field and the gravitational field. The evidence, presented earlier, supports the conclusion that the gravitational effect is electrical in nature, which is fundamental to the establishment of a universal field theory.

It should be noted that another link between gravity and electricity came much earlier. In 1919, Theodor Kaluza of Germany rewrote Einstein's equations in another form. Studying the results, he noticed that one set of field equations were the similar to Maxwell's field equations for electricity. The principles in his paper, "On the Problem of Unification in Physics," were never adopted, and nothing further came of it. This is another example of some of the commonalities of electricity and gravity that have been recognized.

Electricity and magnetism are directly related. The concept of a common _electrical_ field came as a result of the development of Maxwell's equations which define the relationship of the magnetic field to the electrical field. The limited speed of radiating waves was inferred from Maxwell's equations, and the development of the Theory of Relativity was based on these field equations and the Lorentz equation.

Ch. XI - 141

The main problem preventing the resolution of the differences between matter and field is the huge amount of difference in force levels that occur within the atom as opposed to those outside it. Einstein could not rationalize the commonality of the two forces and simply said that matter is an entity in itself. In Chapter IX, evidence was presented that serves to bring these two types of force together in a common equation which represents the rotating electrical field. The main obstacle in forming a unified field theory was thus eliminated by resolving the problem of the incompatibility of the strong force within the atom and the weak force of gravity. The gravity equations of Appendix II show that the extremely strong forces within the atom tend to achieve a degree of balance outside the atom. The electrical balance is not complete, and the itty bitty force that is left over is the force of gravity.

But there is another little stumbling block to the establishments of a universal field theory: the problem of inertia. Einstein reasoned gravitational forces and inertial forces are one and the same since the attraction of gravity was opposed by inertial forces. The possibility that his inertial theory may be incomplete must be considered. Although the two forces are certainly closely related, they are not quite the same.

The hydrogen atom will be used as an example to explain how an inertial force can be created:

> Assume that the hydrogen atom is accelerated by the attraction of another nearby hydrogen atom. Since the electron is rotating around the proton, it will experience the accelerating force first. The accelerating force on the proton will occur somewhat later and will be less than the force on the electron, since it is further away. The reaction of the electron is to move a slightly greater distance than the proton due to the delay in the transmission of forces, thus distorting the electron orbit slightly, thereby resulting in a slight change in the orbital radius and producing a change in the gravitational force. As a result, the reaction force opposes the accelerating force. The same process occurs for the gravitational force, which is why the two forces oppose one another.

Both the inertial force and the gravitational force are determined by the electron orbits of the atoms, and the heavier the atom the greater these forces since larger atoms have more electrons and protons. Are the inertial force and the gravitational force therefore the same? No, because cause always comes before effect since it is the cause that produces the effect. The gravitational force, in this case, presents the cause, and the inertial force is an effect which results in a change in the gravitational force produced by a distortion of the electron orbit.

Although the major difficulties with a unified field theory have now been removed, other secondary problems must also be resolved. The radiation of energy has been described by Maxwell's field equations, but how are these fast moving fields created? The process of electrical conduction was described in Chapter X. Electrons move from one atom to another when activated by an electrical force. The change in the position of the electron produces a local field near each atom, plus an overall field, due to the gross action of a large number of electrons. In this process, the general motion of the electrons is comparatively slow with respect to the speed of electrical waves. Then how is it then possible for an antenna to produce an electrical field which generates waves that travel at the speed of light when the electrons appear to move so *slowly*? This question is not trivial.

While the orbital electron may be moving at a speed near the speed of light, the measurements of current flow indicate that electron speeds are comparatively slow, perhaps due to the random movement of electrons within a conductor. The longer path over which the electrons move, the slower the average velocity along the length of the conductor. This is a clue as to why the length of path of the current is directly related to the wavelength of radiation of an antenna. One possible process by which a fast moving field can be established by slow moving charges is the *group motion* of the electrical charges. It is the speed and distribution of the electrons along the conductor, acting in unison, that produces radio waves which travel at the speed of light. Antennas can be quite large, and it is known that the electrical current in the antenna takes some time to build up. These factors would all tend to support the idea of group action. The following analogy may help to explain the process:

Consider a crowded freeway in Los Angeles with cars moving

very slowly and nearly bumper-to-bumper. The lead car hits a pylon and comes to an abrupt halt. The following cars hit each other, one by one, beginning just after the lead car. Consequently, the point of impact moves down the line of cars at a very fast speed. The shorter the distance between cars, the fast the movement of the point of impact.

The slow-moving electrons in an antenna must act in a similar manner in order to produce waves that move at the speed of light. Because of the necessity for the presence of a driving force to move the electrons, we must also conclude that the speed of which our wave moves has something to do with the speed that an electron can leave its orbit when forced to do so by an external electrical force. Does it take a full revolution of the electron, part of a revolution or several revolutions to leave its orbit? The time that it takes the electron to travel to an adjacent atom must be related to the speed of light since a single photon is believed to be generated by this process. The assertion that the orbital electron is traveling much faster than previously thought will be examined further in Chapter XV.

The major stumbling block, in applying classical theory to the rotating electron of the atom, has been identifying the process by which energy is radiated from an atom. In an electrical coil, oscillating currents produce radiation. The analogy of mass is electrical inductance, and the analogy of a spring is electrical capacitance. Electrical circuits, with currents in a path containing inductance and capacitance, radiate energy in accordance with Maxwell's equations. The electron rotates around the nucleus of an atom, and this is a form of current flow. Why, then, doesn't the rotating electron within the hydrogen atom radiate energy, as might be expected by applying Maxwell's equations to the electrical atom?

Bohr reasoned that atomic radiation occurs only with a change in the electron orbit of the atom, but the method by which an orbital change occurs is not known for certain since the orbital path cannot be easily measured. The answer now seems straightforward. The gravity equations show that the average force on an atom does not change for a stable condition of the atom. In order to radiate or absorb energy, the forces on the atom must vary. Applying a method of reasoning similar to that of Bohr's, and using the gravity equations, the assertion can be made that it is not until the electron changes its orbit that the force changes. Energy is force times distance, which indicates that *a change in orbit produces a change in energy; radiation or absorption.* Electrical circuits radiate energy in accordance with Maxwell's equations, while a stable atom does not radiate. Is the field within the atom thus quite different from external fields, as is the present belief? Why should it be different? The effective current within the atom is similar to an electromagnet which does not radiate energy. It is only when a change in current occurs that radiation is produced. Therefore, by a similar reasoning, the electrons that remain in orbit cannot radiate energy. If the electron alters its path, then the effective current changes, energy is radiated, and the conflict is resolved.

Every force thus far considered is electrical in nature, and so the hypothesis is that the universal field must be an *electrical field*. The view of the universe that is emerging is one which consists solely of an assemblage of tiny, powerful electrical force fields.

The definition of the orbital path of the electron has been shown to be highly important in determining the forces acting on atoms. The problems with the Rutherford model of the atom have been described, and a modified model has been proposed. Orbits can assume any combination of the three dimensions of space, so what is the effect of various three-dimensional orbital paths? The possible orbits of the hydrogen atom and their relationship to radiation characteristics will be examined, but our investigation of gravity is not quite complete. The problems with the gravitational force, that were described in Chapter IV, will first be addressed.

Weighed in the balances and found wanting.

---Bible

CHAPTER XII

The Problems with Gravity

Gravitational theory itself has its problems, as was mentioned in our investigation of the Big Bang theory. Cosmologists are groping with a problem in outer space: galaxies are not behaving properly. Hubble's experiments with Doppler measurements on stars allowed the calculation of the stellar motions from the frequency shifts of their spectrums. Correlating the frequency shift with the estimate of their spatial positions permits astronomers to plot the orbits of the stars. The results are puzzling. The orbital speeds of the stars that are further from the center of the galaxy are moving faster than calculated, as based on their gravitational force of attraction and centrifugal force. These outer stars should be exiting the galaxy due to their high velocities (in accordance with Newton's equation). The gravitational force is calculated as being too weak to hold back the inertial forces, and the stars should be spinning outward from the galaxy.

Another part of the puzzle is the huge cloud of gas, mostly hydrogen and carbon dioxide, believed to be orbiting along with these outer stars. The evidence has led astronomers to believe that there is a tremendous amount of dark matter beyond the outer edge of the galaxy which is affecting the outer stars of galaxies. That is one possibility, and it fits in with other cosmological theories. But if there is so much dark matter in the universe, then how is it that we can see stars so very far out in space with good resolution? Even transparent gases attenuate and distort optical beams because of variations in density. The effect of the heat of summer on a highway produces a glassy image above the surface. The mirage of the desert is another example. This question has been answered by the assertion that the blackness of outer space may be produced by dark matter that <u>completely</u> obstructs the view of many distant stars (is there no intermediate condition?). So the dark matter of outer space depends upon mathematical calculations which are based on gravitational forces. If these calculation contain errors, then such theories must be modified.

The existence of black holes originated from gravitational calculations Astronomers have speculated that every giant spiral galaxy must have a black hole in the center which <u>holds all of its matter in orbit</u>. While cosmologists have asserted that black holes explain certain astronomical events, the evidence to support the theory of black holes is not universally accepted. Nor are all theories consistent. One theory holds that black holes can never be seen since they trap all energy and matter within them. Another theory proposes that black holes can radiate energy during their transition into (or out of) their formation.

Recent images of space from the Hubble telescope show that something is happening at the center of some galaxies, and that action has also been attributed to black holes. If I were an astronomer looking for black holes in space, I would look for nearby stars streaming inward to a point within dark space. Down-shifted Doppler measurements would then tend to support the conclusion that a black hole is present. Black holes, if they exist, would have a significant effect on any masses in the neighborhood due to the effect of their gravitational forces, and these forces would have to also be part of the overall equation used to calculate the movements of galaxies.

What evidence that black holes exist has been found? We do not know the amount of mass within a black hole, and it can only be estimated by its effect on surrounding stars and galactic matter. The presence of fast moving gases circling around a point in space is believed to indicate the existence of a black hole in space due to the gravitational force required to counter the inertial force. Observation of the spectrum of these gases provides information as to their speed, and the shape that they form indicates the direction that they are moving. If a black hole truly exists, then the gases would presumably be moving inward, toward the center of the black hole. Fast moving gases have been observed, and their presence has led to the conclusion that black holes have been observed.

But problems with gravitational theory still persevere. One brave physicist, Mordecai Milgrom, has suggested that our understanding of gravity must be modified for great distances and that dark matter does not exist. Any change in the gravity equation would cause a radical and powerful impact, since it would drastically change our present model of the universe in many ways. There are other reasons for reaching a similar conclusion. It will be shown that a further examination of Newton's gravitational equation, applied to the forces within the earth, leads to some puzzling results.

Before examining the difficulties with Newton's equation, it should be emphasized that there is a great deal of evidence to support it. Lord Cavendish, in 1798, contrived an apparatus containing two large masses and two sets of small suspended masses by which to measure the attraction of each of the two small balls to the large one. The deflection of the small balls produced a reaction force which was used to determine the attractive force between the large and small ball. He expressed the results in terms of the mean density of the earth, which he found to be close to Newton's calculation. The Newton equation has survived scrutiny for many years because of tests like those of Cavendish, and they have (almost) always supported his theory.

In spite of all of the evidence that supports Newton's equation, a problem with the extent of his theory may nevertheless exist. The following question is now posed: "Is the gravitational force between masses the sum of the individual forces on each part of each mass?" The answer to this question is fundamental to the gravity phenomenon. In addition, the "center of gravity" of a mass requires some examination. The following test will be performed: Instead of determining the pull of the earth on a small mass, the pull of the small mass on the earth will be investigated.

The distribution of the attractive forces within the earth have calculated for a small mass of one kilogram placed at a height of ten meters above the earth. The assumption is made that the mass of the earth can be divided into a large collection of small masses, and that each of these individual small masses is attracted to the one kilogram mass with a force given by Newton's equation. The force vs distance in a direction towards the center of the earth are plotted in Figure XII-1. Each of the three curves represents the force per square meter on the earth along a ray from the earth's surface toward the center of the earth at three different angles. Differences in density of the earth have been neglected in these calculations.

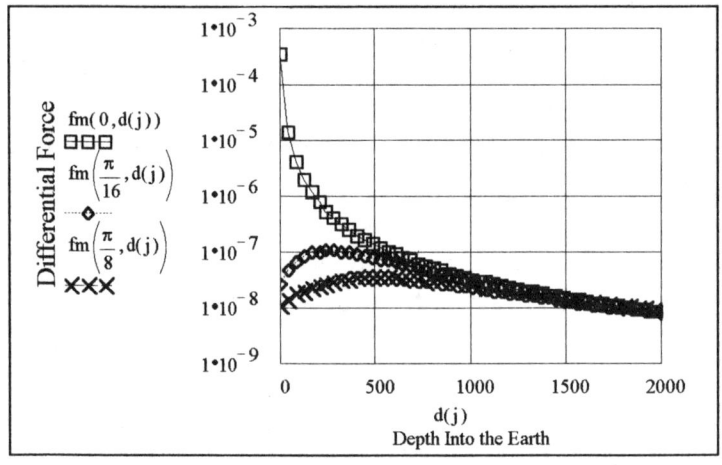

Figure XII-1. Gravitational Pull of an Object on the Earth from the Surface to the Earth's Center at Rays of 0, 11.25, and 22.5 Degrees (height of object = 10 m.)

The gravitational forces on each portion of the earth as produced by the one kilogram mass are clearly far from uniform. As would be expected, the forces near the earth's surface are by far the greatest. The force produced at the center of the earth is essentially negligible, as is also the case at the far side of the earth. The curves of Figure XII-2, which give the sum of the forces along each ray, provide further illustration of the problem. These curves indicate that most of the force is contributed within the first 100 feet of depth, and that the force also decreases rapidly with the angle between the direct ray from the object

Ch. XII - 149

to the center of the earth and the ray at a remote point on the earth. It is necessary to add all of the forces on every ray of the earth in order to have the force equivalent to the weight of the object.

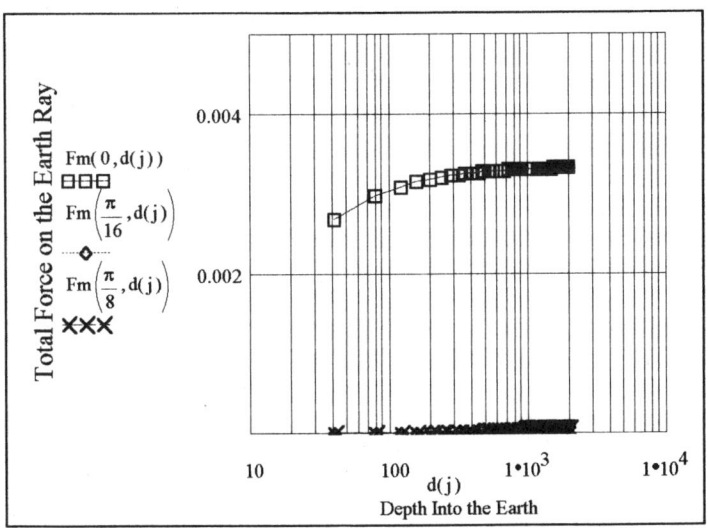

Figure XII-2. Total Radial Forces on the Earth Created by an Object as a Function of the Distance from the Object Towards the Center of the Earth at Three Angles (height - 10 m.)

The gravity equation shows that gravity is obviously a local phenomena for objects located at the earth's surface. Then what happens when the object is raised further above the earth's surface? The height of the object has been raised to 200 meters for the curves of Figures XII-3 and XII-4.

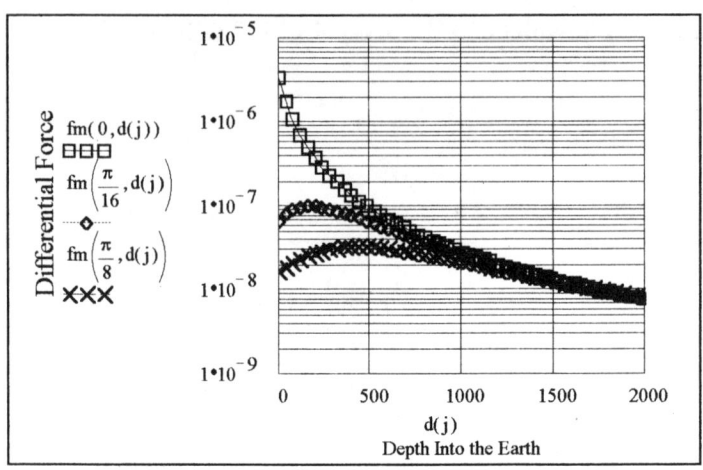

Figure XII-3. Gravitational Pull of an Object on the Earth from the Surface to the Earth's Center at Rays of 0, 11.25, and 22.5 Degrees (height of object = 100 m.)

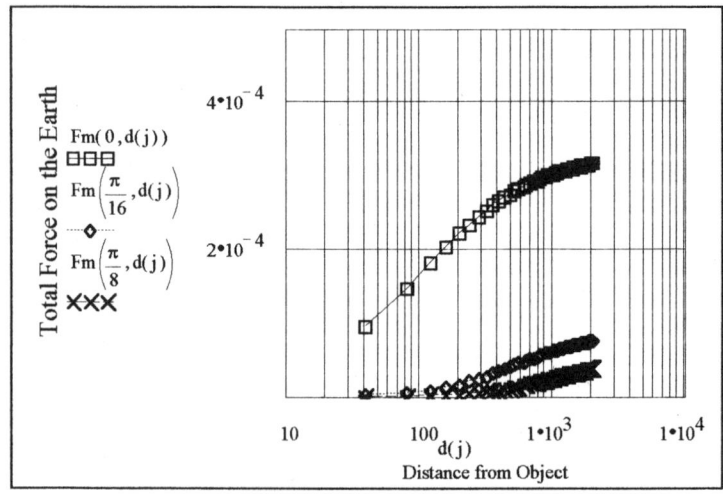

Figure XII-4. Total Radial Forces on the Earth Created by an Object as a Function of the Distance from the Object Towards the Center of the Earth at Three Angles (height - 100 m.)

In Figures XII-3 and XII-4, the forces near the surface of the earth are much lower than those of the first example, as expected, but the shape of the curve has changed. The geometric pattern of gravitational forces on the earth spread out and become less localized as the object is raised to greater heights above the earth, and the attractive forces at the center of the earth and on its far side become more of a factor.

In the above example, the earthly forces act to pull the object in the direction of the center of the earth, as would be expected. However, the center of the area of the gravitational forces, which are exerted upon the earth by the object, are concentrated on areas that are quite close to the surface. The further the object is located from the earth, the more the curve spreads out toward the center of the earth, and each atom and each molecule within the earth begin to exert a more significant effect.

The above analysis leads to a new dilemma. If the gravitational force is the sum of the forces on each portion of a mass, then the above factors must be taken into consideration, in which case the "center of gravity" must not be quite as has been believed. Conversely, if the center of gravity is located at the center of the earth, then the above analysis is incorrect, and something must happen within the mass of an object such that the forces on the atoms within the earth change their mass as a function of their depth below the surface.

Much data has been collected which tends to confirm Newton's equation, which tends to support the latter choice. Milgrom's hypothesis that there is a problem with the gravitational phenomena is more in line with the first possibility. Since the force exerted by each atom within the earth produces nearly the same force on distant objects, the total amount of force exerted on distant stars is higher, thereby having a greater ability to hold them in their galaxy. Consider the following arguments:

1. The sum of the forces on each ray of mass within the earth show that most of the gravitational pull is exerted at the earth's surface for an object located near the surface. At further heights above the earth, the area and volume of masses that contribute to the total force increases significantly. Thus the distributions of forces

from the atoms within the earth changes with height. The forces from all of the rays must be added together to determine the total gravitational pull. With reference to the center of the earth, the variation in force with distance would be expected to exhibit a square-law variation in accordance with Newton's equation.

2. If the gravitational forces are not square law, then a serious problem exists, since either the "gravitational constant" is not a constant and varies with the amount of mass, or else Newton's equation does not accurately describe gravitation. Two balls of equal size and mass weigh the same. When they are placed together they will weigh exactly twice the amount of each ball. The sum of the forces within the earth that are created by these balls must be equal to their combined weight. Calculations are normally made with respect to the "center of gravity" of a body. The center of the earth is far away from an object at the surface. If the measured weight of the two balls does not conform to the forces within the earth as calculated above, then either something is changing when masses come together in aggregates, or else the gravitational concept is in error or incomplete.

3. Newton's Law of Gravity has been well exercised and has never been shown to have any major flaws (although it is incomplete for masses moving at high velocity by Einstein's theory). It has proven to be quite successful in determining the orbits of the planets and their moons. Any modification of Newton's theory will have to conform with these results.

4. It is very difficult to measure the strength of gravitational forces at very large distances. The inertial force can always be mathematically equated to a gravitational force, but if the gravitational constant is

really a variable, then errors in the calculations are possible.

5. If the gravitational constant proves to be a variable, then forces that are further away from the surface of an object become more significant as the distance between the two objects increase. Greater forces, exerted at further distances, may solve the galaxy problem mentioned above.

Resolving the above problems could lead to a better understanding of gravity. The following comments apply to the example given above:

1. These equations show that the inner core of a large body has very little effect on the gravitational force exerted on objects located nearby (by applying Newton's equation to the individual masses that make up the whole). The earth could just as well be hollow, and the gravitational force on an object on the surface would not change appreciably. Even if the core of the earth were far more dense than the surface layer, then the weight of objects on the surfaces would not exhibit much variation.

2. According to the above analysis, far distant objects, such as stars and planets, exert a more significant effect on the gravitational forces at the center of the earth as compared to closer objects.

3. The sum of all of the forces of all of the rays toward the center of the earth were not calculated. It is still possible that Newton's equation will hold since the spreading effect tends to counter the concentration of forces near the surface of the earth. Should a conflict be found to exist, then the characteristics of atoms must somehow change as a with the amount of mass and of their location within the mass. Such a result would be

profoundly significant.

The above analysis illustrates that the present gravitational model may be incomplete. Variations in the gravitational force at the earth's surface have been measured and are well known. Such variations have been attributed to a variable gravitational field, cause unknown. On the basis of the above analysis, it is likely that they are caused by local variations in the mass of underground substances that lie quite near the earth's surface.

None of these comments conflict with the analysis of the gravitational force that was presented in Chapter IX. The difficulties described above would lead to the belief that the gravitational model requires further development in order to resolve all such problems. While a full gravitational model requires further investigations of this type, they are beyond the scope of effort of this book.

Non sequitur [it does not follow]

---

*My heart leaps up when I behold
a rainbow in the sky.*

---Wordsworth

CHAPTER XIII

What Produces Atomic Spectra?

Our human eyes can only see the visible wavelengths (colors) of radiation. A rainbow in the sky produces most of the colors that we can see. A much wider range of wavelengths (spectrum) is constantly invading our atmosphere from space. The radiation of light from stars in the sky has provided highly valuable information from which the structure the universe can be deduced.

Radiation from space is presumed to be produced by the excitation of atoms in the masses of stars (planets merely reflect starlight). While the velocities of the movements of stars can be calculated on the basis of the Doppler shift, the method of generation of an atomic spectrum is much less clear. The characteristics of the discrete frequency spectrum of energy radiating from atoms has long puzzled scientists, since it does not resemble any other known spectral patterns. Classical analysis has not produced great success, while the methods of quantum physics have simulated spectral patterns of certain types of radiation with fairly accurate results. However, the use of energy concepts, upon which quantum physics is based, may not lead to a full comprehension of what is happening within the atom. While the Law of Conservation of Energy provides a firm basis upon which to form conclusions, it is *force* that governs the universe, and ambiguities can result when only energy concepts are utilized. The importance of electron orbits of atoms has been emphasized in earlier chapters. A better understanding of the method by which atomic spectra are generated is sought, and an attempt will now be made to determine how the spectrum of radiation from an atom is created.

Hydrogen, the most simple atom, has been the subject of

numerous investigations of its distinctive radiation particular pattern of sharp spectral lines that correspond to wavelength (frequency). The other known types of radiation, that do not exhibit the discrete patterns of radiation from atoms, will be described briefly for comparison purposes.

The "black body" energy radiation, that has been previously discussed, is produced by mass in the process of losing heat. As a black body heats up, it radiates energy at various frequencies/wavelengths over a frequency band without any gaps (missing frequencies), and, as its temperature increases, the curve shifts towards lower wavelengths (higher frequencies). The Wien mathematical model of blackbody radiation produces a graph related to the distribution of the speeds of molecules in a gas (the "distribution law"). By equating the energy distribution of radiation to the distribution of the energies of gas molecules, Wien arrived at the formula for the rate of energy emission. This formula can be used in various types of mathematical analysis, such as predicting heat absorption. The Planck model is somewhat more accurate, and is used to predict the expected various energy levels of the moving molecules.

A black body has no gaps in its radiation spectrum. A continuous spectrum implies that the radiation is not harmonic and is associated with random movements (a discrete spectrum has gaps and is related to harmonic motion). Random movements of atoms will produce collisions, which suggests that heat radiation and molecular movement are related. Not all heat radiators are black bodies. Radiation and absorption of optical energy depend on how "shiny" the surface may be. A shiny surface reflects and absorbs less energy than a dull surface, and a duller surface has less "emissivity." The "emissivity factor" of the surface of the radiator accounts for the additional differences in the radiation of a mass at elevated temperature. At absolute zero, there is no temperature radiation since the structure of the mass does not vibrate.

Another type of radiation is produced by radioactive atoms. Most radioactive materials are heavier than lead and can generate alpha, beta, and gamma rays. Radioactive radiation is of much higher energy than optical radiation. One such material is radium which generates high-energy alpha particles, plus some beta and gamma rays from its decay products. Only one gram of radium releases 140 calories of heat per hour, which is a surprisingly high amount. Since one calorie is sufficient to raise the temperature of one gram of water 1°C, one gram of radium, placed in one gram of water, would cause it to boil within less than an hour. The radium constantly radiates energy, and Einstein proposed that mass and energy are equivalent. Therefore, the mass of radium is diminished by radiation, but by a very small amount, indicating that it does not take much mass to create high levels of energy. Therefore, radioactivity can continue for a very long period of time.

Manmade radiation has unique characteristics. Radio transmissions produce several types of radiation patterns, which depend upon the method of placing information on the channel frequency. For amplitude modulation (AM), the amplitude of the carrier frequency is modulated (the energy level changes), usually by speech. For frequency modulation (FM), the frequency is modulated. Digital or pulse modulation varies the characteristics of the pulse and/or the repetition rate. However, the only resemblance of these spectral shapes to that of atomic radiation is that the spectral lines are sharp (discrete).

Atomic radiation, produced by the excitation of atoms, has gaps in its radiation spectrum, and the spectral patterns are comparatively fixed and recognizable by the location of the lines of the spectrum. The generation of a discrete spectrum requires harmonic (cyclic) motion. The mechanism by which this occurs is complex and difficult to analyze. The early methods of predicting the spectrum of atomic radiation were based on classical analysis, but they proved to be unsuccessful. The reason for the failure is that improper analogies and assumptions were used. The electron was considered to be a small particle in orbit around the nucleus of the atom; in order to obtain radiation the electron would have to fall continuously towards the nucleus -- a distinct impossibility. Also, a rotating electron constitutes an electric current moving in a circular path, and using analogies with electrical circuits, the unexcited atom would then be expected to radiate energy.

A plausible explanation was finally obtained by Bohr. If the radius of the atom was allowed to assume any size, his mathematical equation could be adjusted to conform with the radiation spectra. Formulas, based on his principle, produced results that correlated exactly to the frequency spectrum. The success of these methods led to their acceptance. As a result, the realistic classical analysis of Planck evolved into the energy concepts of quantum physics, as based on Bohr's hypothesis.

Today's quantum physics model of the hydrogen atom pictures an energy barrier at the center of the atom that prevents the electron from falling into the nucleus, while outside of the barrier the electron is attracted to the nucleus. No reasons are given for the existence of such a barrier, and most calculations are based on energy aspects. Since energy is always conserved (except for radioactivity which alters mass), the gross errors and mistakes in the analysis are less likely. On the other hand, this method does not provide a good picture of exactly what is going on inside the atom since it cannot predict the electron orbit directly as would be possible with classical analysis. Since classical analysis has not been very successful in producing accurate results, quantum physics is believed to be the only useable method that can predict the characteristics of atomic radiation. Quantum methods have predicted orbital paths of the electron of the atom which pass directly through the nucleus. But the electron cannot pass through the nucleus, since it is believed to have an energy barrier which repels the electron, and therefore the theory has contradictions.

In order to apply classical analysis to the determination of the spectrum of hydrogen, it is necessary to use nonlinear analysis since the electrical force between charges is nonlinear. The major failing of classical analysis can be eliminated by using the Bohr assumption that harmonic states exist for the atom wherein no energy is lost for a fixed atomic radius. Bohr's idea of a huge orbital radius must, however, be eliminated in order to maintain realism. The methods of nonlinear analysis can be very difficult, and sometimes so complex that deriving solutions is a nearly impossible task. The mathematics of nonlinear analysis is beyond the scope of this book, although a brief verbal description of the method of choice is presented:

> The forces acting on an atom are nonlinear. A standard form for a second-order nonlinear equation with a periodic coefficient is called *Hill's differential equation*, a nonlinear differential equation which has a well known format. If Maxwell's equations are applied to the atomic electrical charges, forces, and velocities, the frequency of the electron in its orbit is found

to be inversely proportional to the radius. Assuming only a small percentage change in radius, a Hill's equation can be formulated from these parameters. If the periodic coefficient of the differential equation is a cosine function, then this specific form of Hill's equation is called a *Mathiew equation*. The Mathieu equation is a much-studied nonlinear equation which can have both discrete stable solutions and unstable solutions. A radius that grows without bound would be considered an unstable solution that would reflect ionization (separation of the electron from the nucleus) of the hydrogen atom. The amount that the radius changes, in order to produce ionization, is not great, since the electron need only pass to the next nearby atom. A stable solution produces a discrete frequency spectrum that is produced by parametric frequency variations. There are many types of solutions to a Mathieu equation (some of them very complex), depending on the variable coefficients. For example, a dissipationless electrical circuit, with a varying capacitance, (frequency modulator) has stable and unstable areas of the frequency spectrum. By applying the Mathieu equation to atomic systems, it is possible that the atomic frequency spectrum of radiation can be predicted with accuracy.

Radiation can be generated in many ways, and one method is to generate radiation signals using electrical circuits. The resulting radiation is in the form of specific patterns that can be related to the mechanism by which it is created. The spectrum of hydrogen, however, takes a form that is unfamiliar, not resembling that of any radiation created by electrical circuits. The assertion, made here, is that the unusual spectrum of hydrogen is produced by a nonlinear system action, from within the atom, when an exchange of energy occurs. Most systems in nature are nonlinear in way or another, but, for ease of analysis, they are approximated as linear systems. The complex nonlinear Mathieu equation is more versatile and has a variety of solutions, depending upon the coefficients of the nonlinear equation. Some of these solutions produce some very unusual spectral shapes. It is known that a nonlinear system, such as a nonlinear resonant circuit, can distort the resonance curves by bending them as shown in Figure

Ch. XIII - 160

XIII-1. The bending of the spline of the curve produces the distorted frequency response curve of Figure XIII-2.

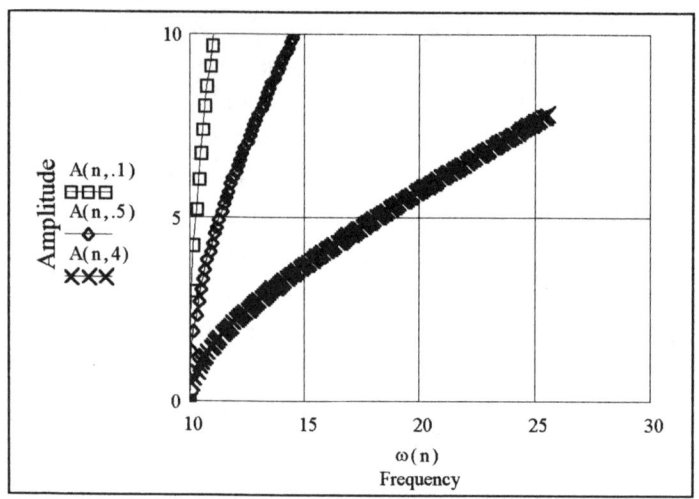

Figure XIII-1. The Curved Spline of the Frequency Response of a Nonlinear System.

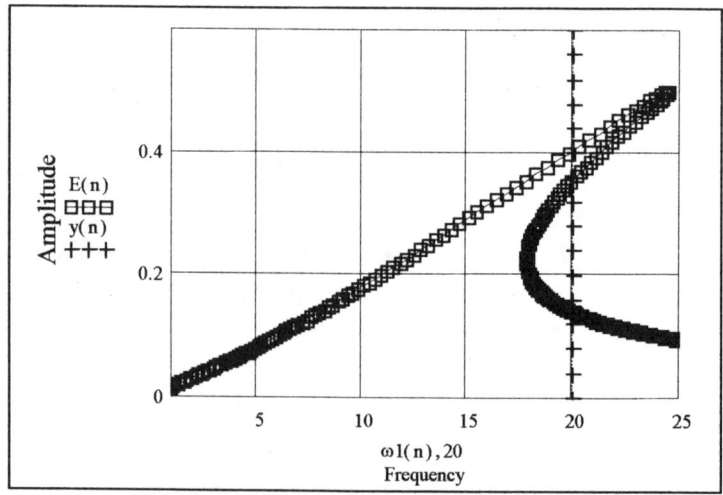

Figure XIII-2. Frequency Response of a Nonlinear System (the vertical line is in the unstable region)

This system has an unstable region on the right side of the curve. The sloping region to the right fits the requirement for a discrete change in energy, since vertical level jumps are thus possible. Most man-made radiation produces a symmetrical spectrum, while atomic radiation is usually asymmetrical. The bending of the curve accounts for the asymmetry. Allowable frequencies/wavelengths, produced within the unstable region to the right, depend upon the geometry of the electron orbit. Each traverse of the orbital shape must be repetitious in order to produce radiation spectral lines. The spectrum of hydrogen can be modeled by correlating the electron orbits with the measured radiation frequencies. Once the orbital paths are determined, the knowledge of the radiation process will be known. It is therefore possible to apply classical analysis to a nonlinear system in order to determine the energy spectrum of hydrogen.

The derivation of the nonlinear hydrogen (differential) equation will not be attempted here, but all of the ingredients that can be used to make the necessary determinations are available using the Mathieu equation. The hydrogen atom is a nonlinear system due to the fact that the forces on atoms vary with the square of the distance between electrical charges. Although it has proved to be a convenient artifice in quantum theory, it is not necessary to invent an impenetrable wall near the nucleus of the atom to form a system model that produces a suitable spectrum. The force on the electron can be parametrically modulated by either an applied electrical field or a photon. With such techniques, a Mathieu equation can be developed in order to develop an accurate classical model of the hydrogen atom.

In the Mathieu equation referred to above, the frequency variation is represented by jumps in the radius of the electron orbit, and variations in path configurations are therefore possible. The radiation of energy is directly associated with these jumps. The radiation does not, however, occur in a 1:1 correlation with the orbit, but rather with the *change in orbit*. Therefore, the part of the orbit that does not change cannot be sensed since it does not produce energy radiation. It is therefore impossible to perform an inverse transform of the spectrum to calculate the exact orbit, since there can be many possible geometries which are harmonic but do not change with time. Using these methods, it may be possible to eliminate an important contradiction of quantum physics: an electron orbit passing through the nucleus of the atom.

The above analysis leads to a different approach to the problem. Instead of using the radiated spectrum to derive the change in the electron orbit, it is possible to use the reverse approach and determine the spectrum on the basis of electron orbits. This method requires the construction of an electron orbit which can then be altered by modulating them in various ways, thereby forcing a change in orbit with some form of input energy. Since the electron orbit can pass through the three dimensions of space, more than one type of orbit can produce the same (or similar) spectrum, but with a different spatial radiation distribution. This approach will be time-consuming, however, since many potential orbital paths must be examined in order to match the calculated spectrum with the measured spectrum.

The above logic implies that it is not impossible to find classical methods to analyze atomic spectra. The difficulties with nonlinear analysis should not be dismissed, however. The calculations can be exceedingly complex. The ready availability of powerful computing tools becoming easily available it will provide the capability of performing the proper nonlinear analysis.

We have seen much evidence to support the belief that the orbital path of the electron is of great importance. The derivation of some of the possible electron orbitals of the rotating electron is the next logical step for analysis.

Curved is the Line of Beauty.

---W. MacCall

CHAPTER XIV

Atomic Orbitals

It is not yet possible to obtain a good image of the interior of an atom with current laboratory equipment. Scientists have performed special measurements that can be used to estimate the possible electron orbits. One method is to shoot energy waves, such as coherent light, through a thin section of solid material and then analyze the resulting exiting energy patterns in order to deduce the orbital patterns. Unfortunately, the analysis is complicated and inaccurate because the particle must pass through several atoms before exiting and only the combined effects are available in the data. Another method is to analyze the spectral characteristics of the radiation exuding from the atoms when they are excited by an external energy source, such as an electron beam. This method also has its pitfalls.

The first problem is that the *mapping* from one set of variables to another is not always one-to-one. Mapping can be described by an example. Consider a three-dimensional map of the entire surface of the earth on a sheet of paper. The three dimensions of space have been transformed to two dimensions on the map. If the mapping is 1:1, the earth is accurately represented on the paper, and a unique three-dimensional representation of the earth is possible from the paper map. Otherwise, more than one volumetric shape can be formed, and the mapping is not 1:1.

A 1:1 mapping does not always occur in physics applications. Consider, for instance, a wave generated in a body of water. There are several ways to generate waves, and the characteristics of the wave do not necessarily indicate the mechanism by which the wave is generated, since it is possible for different actuating movements to produce wave patterns which resemble one another. Waves can also be generated by other actuators such as wind. Thus a 1:1 mapping of the waves and the movements at the actuating source may not be possible. A similar problem exists in electronics since the spectrum FM modulation of a radio transmitter, at a low modulation index, can resemble that of AM modulation, and under these conditions parametric modulation and power modulation can produce similar spectra.

It is even more difficult to analyze the waves generated by atoms, since a group of atoms can be considered as a mechanical or electrical system with several degrees of freedom. For instance, the atom exists in three physical dimensions, and motions can be produced in several planes with more than one atom either acting in concert with others, or by itself, to produce similar waves. The analysis of electrical networks shows that more than one type of circuit can produce a given response for a given excitation unless all of the possible responses for each degree of freedom are considered. The computational problem can therefore become extremely complex.

The analysis of the spectrum of radiation of atoms is accomplished in quantum physics theory by utilizing the known physical characteristics of energy. The total energy of a system consists of the static energies plus the dynamic energies, and energy is always conserved. Unfortunately, the use of energy concepts does not necessarily guarantee 1:1 mapping, which is one of the main problems with this type of analysis. A more exact analysis is possible for the harmonic excitation characteristics of a crystalline structure wherein all of the atoms begin to vibrate in unison, and certain reasonable conclusions can thus be drawn:

Since the mapping for the structure of mass from its radiation spectrum is not usually 1:1, we have another reason for suspecting that the atomic orbits of the electrons, as predicted by quantum physics, are subject to question.

It was asserted earlier that energy radiation corresponds with the *change* in the path of the orbit rather than the orbit proper. Such an assumption is reasonable since the atoms do not radiate energy unless excited by some form of energy such as temperature or light, thereby affecting the position of the electron in its orbit. Therefore, the exact orbit is very difficult to portray, since only information produced by the change in orbit can be detected. A somewhat different approach is therefore required.

Classical analysis can again provide the desired answers, since stable conditions, wherein energy changes do not occur, can also be described. Nonlinear control system analysis, as applied to the atomic model, shows that there are certain stable orbital paths (which require specific limits for the coefficients of the nonlinear differential equation). Using an analogy of the orbiting electron with an electrical system, we can consider it to be a lossless oscillator with two (dimensional) degrees of freedom (there are really three degrees of freedom due to the three dimensions of space, but considering all three makes the calculations very difficult, and the two-dimensional system is used for simplicity). The calculations indicate that the percentage change in orbit associated with energy radiation is quite small. This result is acceptable, since the force between the electron and the nucleus is inversely proportional to distance, and the amount of energy radiated decreases with an increase in radius, thus correlating with measured data.

Unfortunately, nonlinear analysis is very complex, and a detailed analysis of this type is beyond the scope of this book. A more simple method of determining the orbit is desired. Performing the inverse transform of the spectrum obviously does not produce the proper orbit. Of one thing we can be certain: energy radiation and the electron orbit of an atom are directly related. The approach will be to assume an orbit and calculate the energy spectrum that is generated by a change in orbit. Selecting an orbital pattern does not necessarily guarantee that the correct orbit is obtained, since the mapping may not be 1:1, but if the spectrum correlates with the measurements, then the orbit qualifies as a potentially correct pattern.

This approach appears to be similar to the present methods of quantum physics, but there are significant advantages. When an atom is excited by an external energy source, its orbit can change in many ways. Space has three dimensions, and the orientations of the radiation source and the atom affect the results. The type of modulation is also related to the change the force, the energy, or the parameters of the system, and each of these modulation methods can produced different spectral patterns.

Several of the above methods of excitation have been investigated. Orbital paths over three dimensions can have complex patterns. Two computer simulations programs were used for the analysis. The first set of orbits was computed using the *Mathcad* computer program (*Mathcad* is a registered trademark of MathWorks, Inc.). Figures XIV-1 through XIV-9 illustrate the envelope of the path of the orbital electron when the radius of the orbital path is modulated by a sine wave. The mathematical descriptions of the orbits are given in the appendix. The amount of modulation and its direction in space produce a considerable change in the three-dimensional patterns. The first three curves have different levels and orientations of orbital modulation.

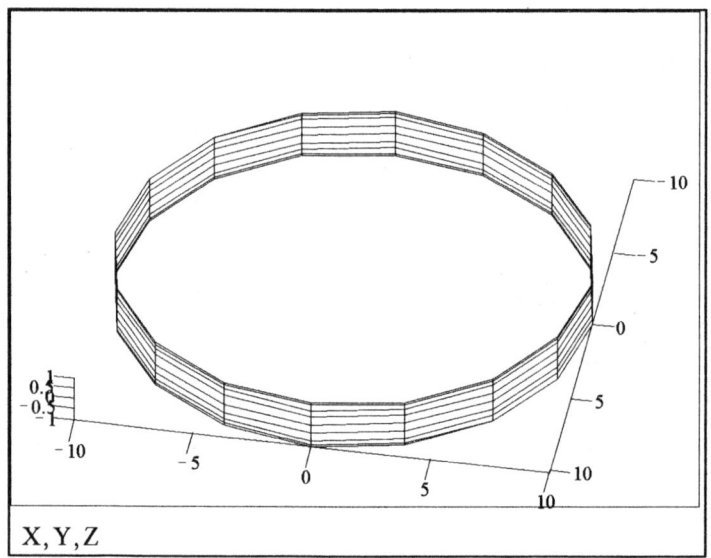

Figure XIV-1. The Envelope of the Path of the Electron of the Hydrogen Atom for Spatial Modulation Along the Z-Axis

The exact orbit is not plotted using this program, and the electron is moving up and down as it rotates around the nucleus of the atom.

Ch. XIV - 167

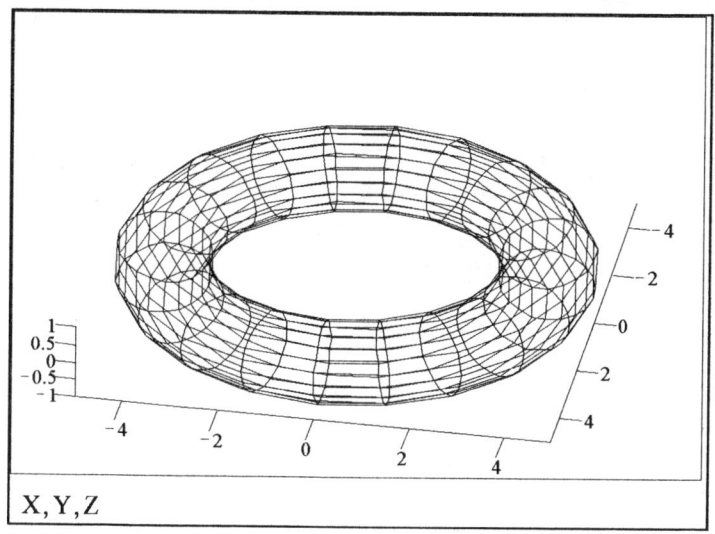

Figure XIV-2. The Envelope of the Path of the Orbital Electron of the Hydrogen Atom for Spatial Modulation in Three Planes

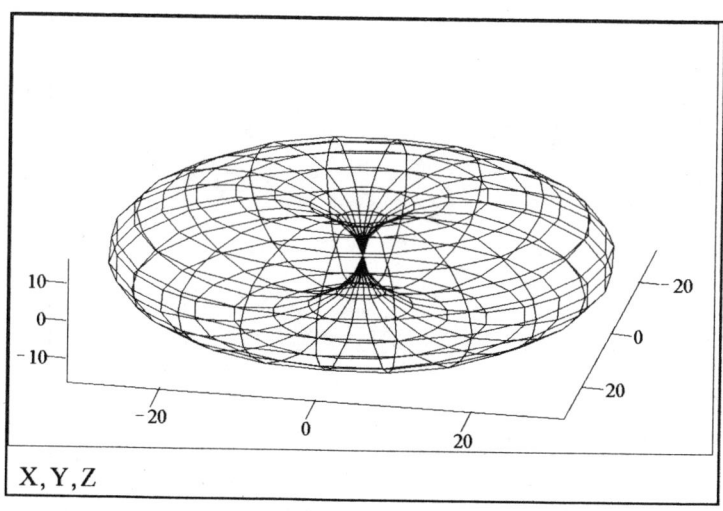

Figure XIV-3. The Envelope of the Path of the Orbital Electron of the Hydrogen Atom for Spatial Modulation in Three Planes.

Ch. XIV - 168

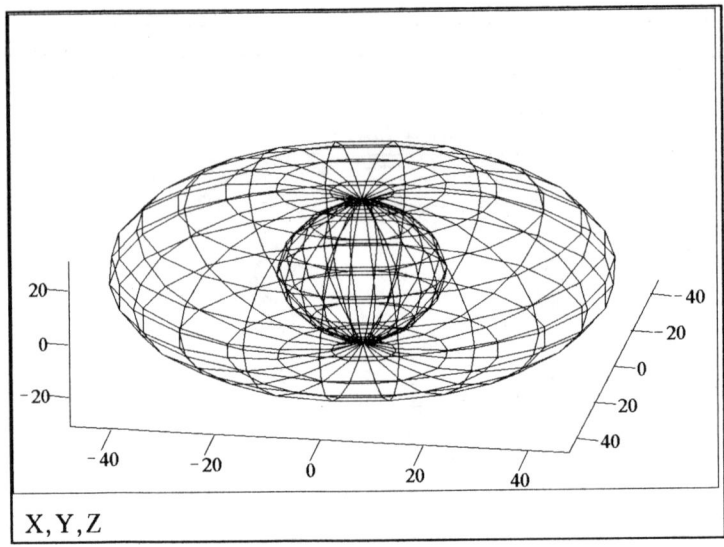

Figure XIV-4. The Envelope of the Path of the Orbital Electron of the Hydrogen Atom for Spatial Modulation in Three Planes.

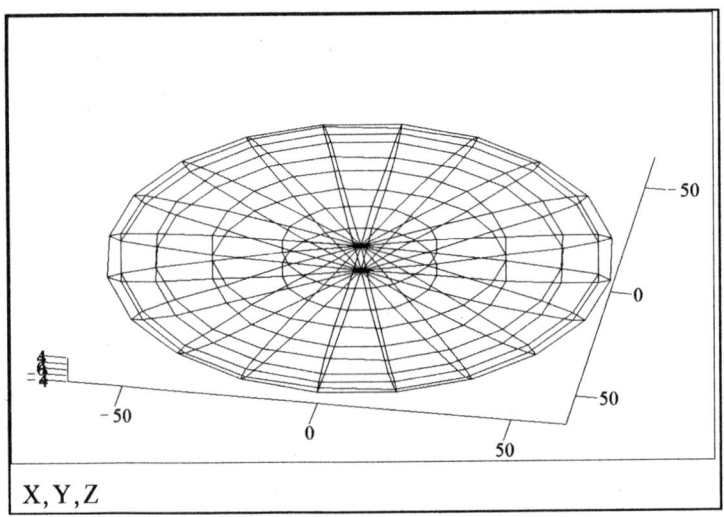

Figure XIV-5. Envelope of the Path of the Electron of the Hydrogen Atom for Spatial Modulation in Three Planes

Ch. XIV - 169

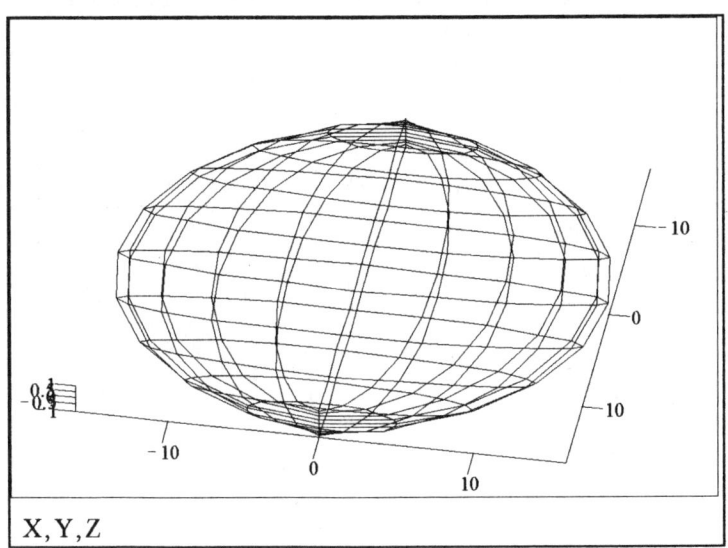

Figure XIV-6. The Envelope of the Path of the Orbital Electron of the Hydrogen Atom for Spatial Modulation in the X, Y and Z Planes

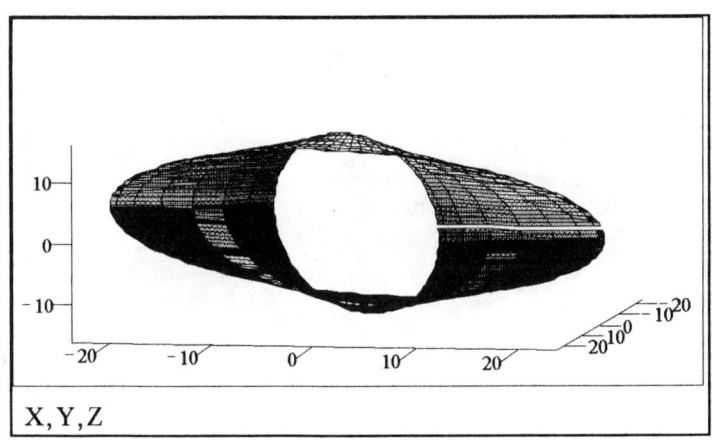

Figure XIV-7. The Envelope of the Path of the Electron of the Hydrogen Atom for Spatial Modulation in the X and Y Planes

The movement of the electron (in the x-direction) of Figure XIV-7 is shown in Figure XIV-8. A similar variation occurs along the y-axis.

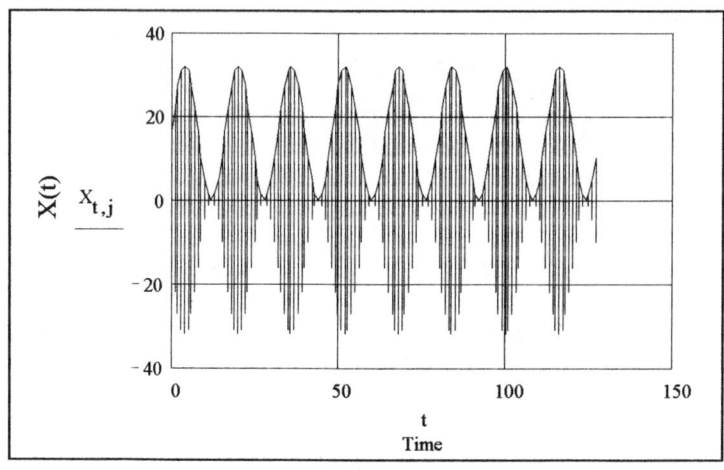

Figure XIV-8. The Variation of the Path of the Orbital Electron
of the Hydrogen Atom in the X-Direction

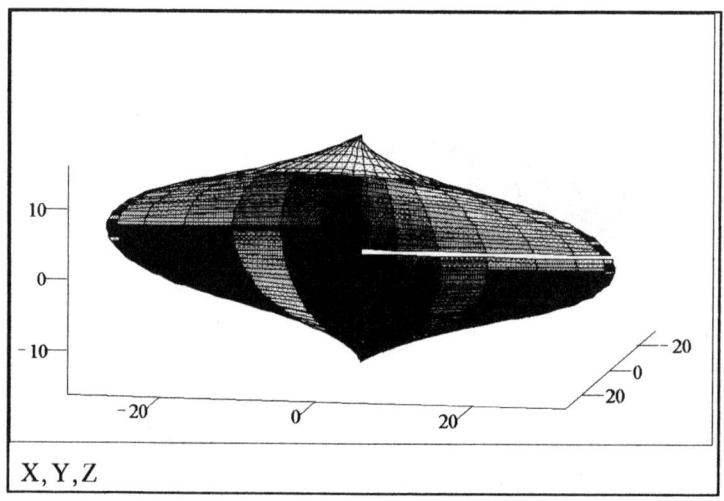

Figure XIV-9. The Envelope of the Path of the Electron
of the Hydrogen Atom for Spatial
Modulation in the X and Y Planes

Simulations of this type were also plotted in Figures XIV-10 through

Ch. XIV - 171

XIV-13 using the *Matlab* computer program (*MATLAB* is a trademark of The MathWorks, Inc.).

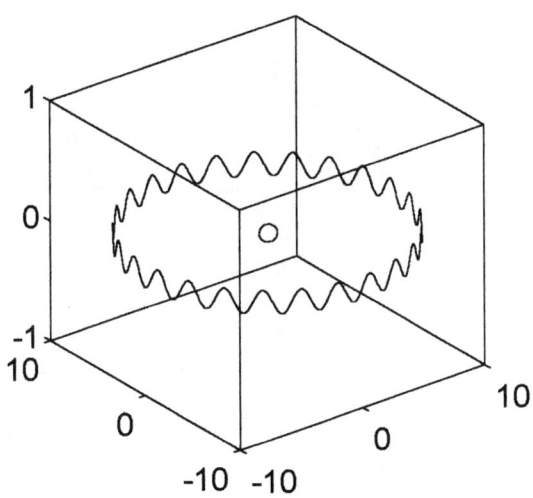

Figure XIV-10. Spatial Locus of the Hydrogen Orbital Electron Modulated in One Plane

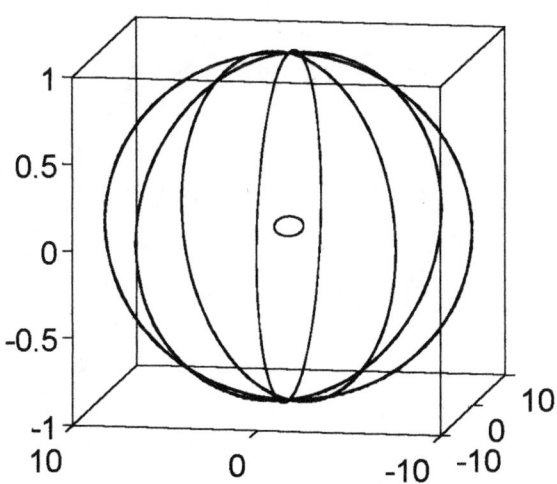

Figure XIV-11. Spatial Locus of the Hydrogen Electron Orbit Modulated in Two Planes

Ch. XIV - 172

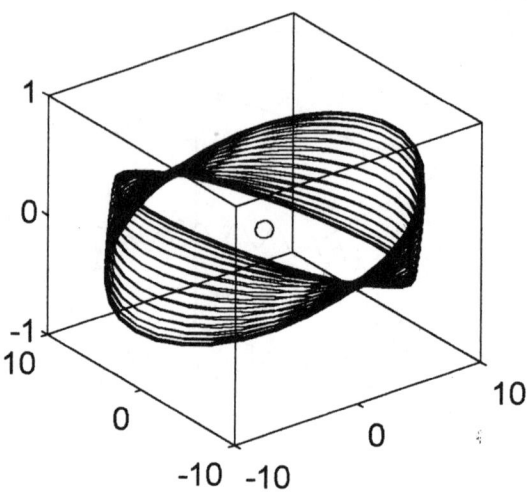

Figure XIV-12. Spatial Locus of the Hydrogen Orbital
Electron Modulated in Two Planes

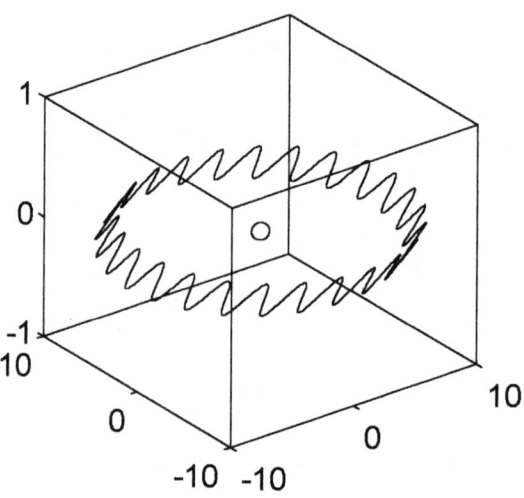

Figure XIV-13. Spatial Locus of the Hydrogen Orbital
Electron Modulated in One Plane

Ch. XIV - 173

The latter set of graphs picture the exact path of the electron as it travels around the nucleus of the hydrogen atom, while the earlier graphs showed the envelope of the orbital locus.

Figures XIV-14 and XIV-16 show orbital shapes that produce spectral lines which are similar to that produced by energy shifts that occur for an atom that is in an electric field (called the Stark effect). The radiation frequency spectra of these orbital patterns are shown in Figures XIV-15 and XIV-17, respectively. Note that the line splitting occurs in two planes.

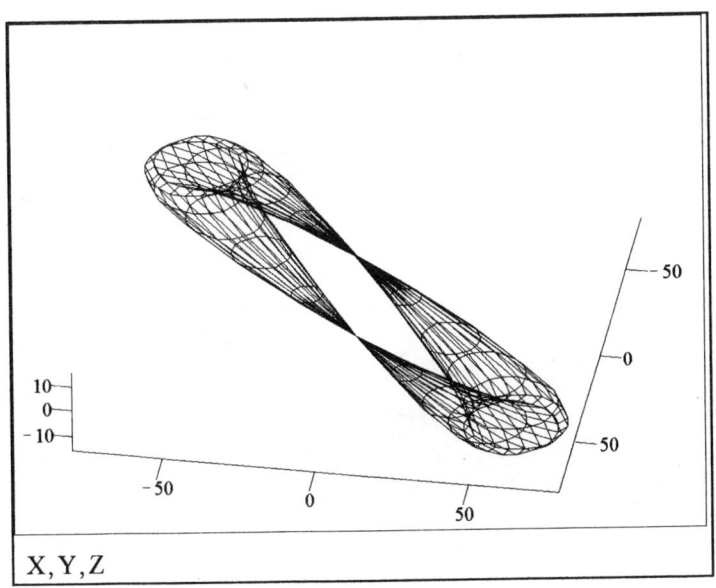

Figure XIV-14. The Locus of the Path of the Electron of the Hydrogen Atom for Spatial Modulation in Two Planes

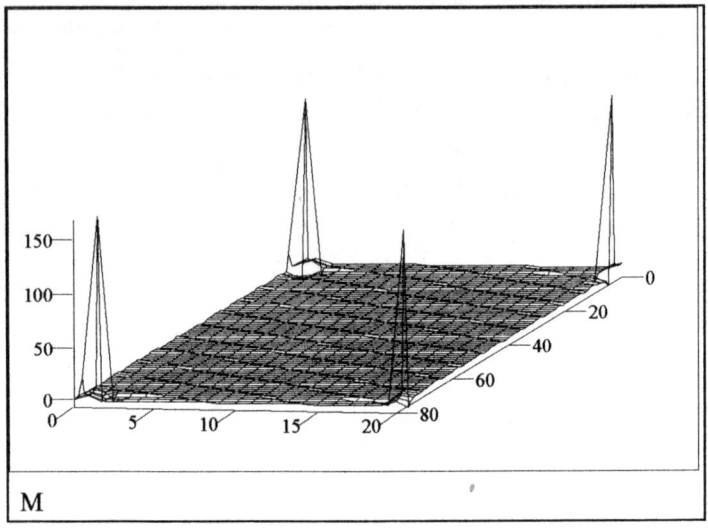

Figure XIV-15. Spectral Lines of the Orbit of Figure XIV-14

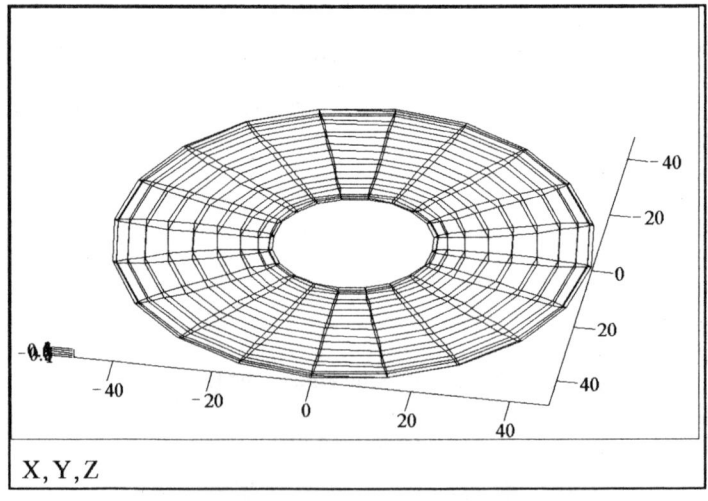

Figure XIV-16. The Locus of the Path of the Electron of the Hydrogen Atom for Spatial Modulation in Three Planes

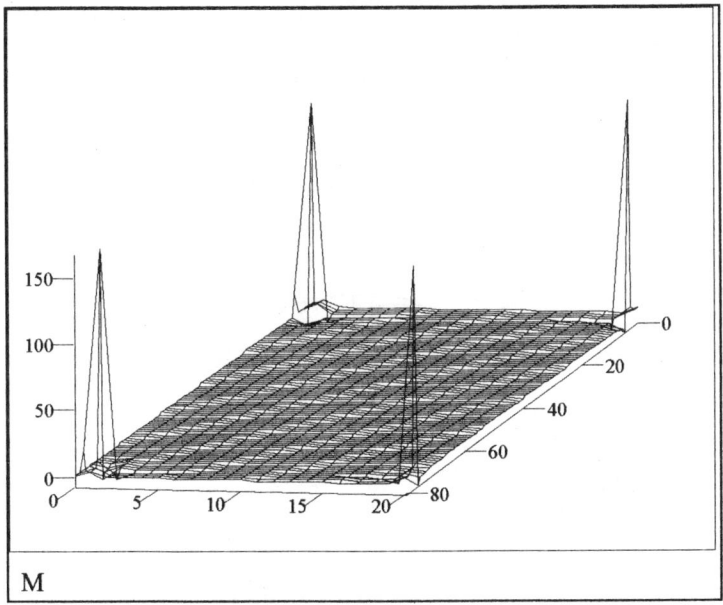

Figure XIV-17. Spectral Lines of the Hydrogen Atom
for Electron Modulation in Three Planes

The above orbital patterns can be correlated to the modified Bohr model of the atom. Instead of having an arbitrarily large orbit, some of the above orbits, such as that of Figure XIV-12, have several passes around the nucleus of the atom for each cycle, and the path <u>length</u> per cycle can attain any length, depending upon the number of passes. In other words, the radius need not be huge per the Bohr hypothesis, but the *path length* can become quite long by altering the path as shown in the method given above. Using models of this type, the construction of an atom in space. can be pictured

An interesting pattern is obtained when the electron is parametrically modulated. In Figure XIV-18 the angular rate is phase-modulated (moving the angular position of the electron in its orbit). Note the little swirls in the pattern. Is it possible that photons are created in this manner? Photon waves must be quite small in size, since they have the capability to knock an electron out of orbit, while long

Ch. XIV - 176

waves are not know to produce electron ejection.

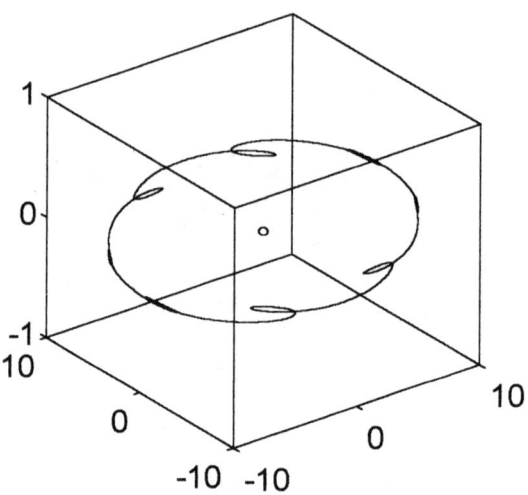

Figure XIV-18. The Spatial Locus of the Electron Orbit of the Hydrogen Atom With Parametric Modulation of the Electron

While these are orbital patterns that we might suspect are present within the atom, most of these spectral patterns do not correspond to the radiation patterns of hydrogen. The frequency spectrum of Figure XIV-17 resembles the spectrum produced by the Stark effect. However, none of the orbital patterns thus far considered produce spectra which resemble that of hydrogen atoms which are excited by radiation. Obviously, the orbital patterns must have a different shape in order to produce the characteristic spectrum of hydrogen. The simulation of the spectrum of a frequency-modulated electron is shown in Figure XIV-19, and it has a spectral pattern that some resemblance to that of measured atomic spectra. In this case, the frequency of the moving electron is modulated by a *circular function*.

Obviously, other orbitals produce different spectral patterns, and so proper alteration of the orbital path may result in patterns which produce spectra that more closely resemble that of the hydrogen spectrum. The Bohr "radius" requirement can also be met using other patterns, and it should be possible to obtain a Fourier spectrum that is close to that of hydrogen by manipulating the orbital equations.

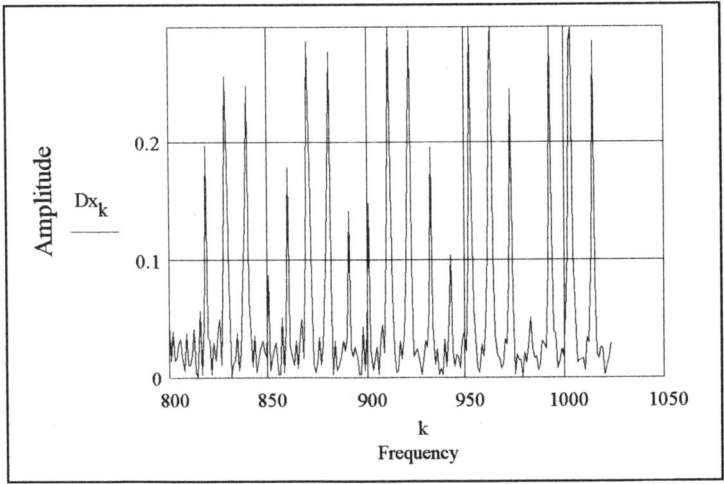

Figure XIV-19. Frequency Spectrum of a Hydrogen Atom With the Orbital Movement Frequency-Modulated in Two Planes.

A crystal specimen is three-dimensional, and all possible geometric forms of excitation must also be considered when a <u>molecule</u> is not symmetric. For instance, calcite has the property of "birefringence" and produces a double image when light passes through it. When a crystal possesses double refraction, it means that the time that it takes for radiation that passes through the calcite crystal varies from plane to plane, passing through the crystal along two paths. Why does it slow down more along one path rather than another? It must be due to the asymmetry of the molecule, but many molecules lack symmetry and are not birefringent. The most likely possibility is that the electrons must take a longer time to change orbit in one plane than another. Again, the orbital path comes into play.

Classical analysis of this type can lend much greater depth to the understanding of atomic physics, since it produces a picture of what is happening within atoms. A more accurate model of the atom may result by adding supplementing quantum methods with analyses of this type.

The causes of events are ever more interesting than the events themselves.

---Cicero

CHAPTER XV

The Speed of the Electron in Orbit

The speed of the electron in its orbit around the nucleus of an atom has not been measured directly, and is determined by calculations. In order to make a determination, the electron is first assumed to be a particle. Using the accepted method of substituting Coulomb's law for force into Newton's equation relating force and mass, the velocity of the electron in the hydrogen orbit is calculated to be

$$v_h = 2.23 \times 10^6 \text{ m/sec.} \qquad [\text{XV-1}]$$

This velocity is a little less than 1% of the speed of light.

The premise that an electron is a particle is questionable, since it is known that an electron sometimes acts as a wave. In an earlier chapter, the assertion was made that the electron must be traveling at a much higher speed. Other methods of estimating the speed of the electron in its orbit will now be presented which are more compatible with a universal field, .

Maxwell's electrical field equations are applicable to both radiating fields and confined fields. Applying these equations, the speed of light varies with the electrical constants of free space, μ_0 and ε_0 (permeability and permittivity), which relate to the electrical parameters of inductance and capacitance. It would therefore seem reasonable that the inductance of the orbital path and the electrical capacitance between the electron and the proton are the limiting factors for the speed of the electron in its orbit. The speed of the orbital electron will be calculated using these parameters.

Ch. XV - 180

The electron orbital is analyzed as an electrical circuit having inductance and capacitance. The self-inductance of the orbital loop of hydrogen is proportional to the path length of the electron orbit,

$$L_h = 1.66 \times 10^{-17} \text{ Henry}. \qquad [XV-2]$$

The calculation of capacitance is defined in terms of the area and separation of charges, and then multiplied by the permittivity constant. There are <u>two ways</u> to calculate capacitance. Using the area of the orbit, and the separation between the electron and the proton, to calculate the capacitance,

$$C_h = 1.476 \times 10^{-21} \text{ Farad}. \qquad [XV-3]$$

With the other definition of capacitance, as the electric charge divided by the voltage, we get

$$C_h = 2.95 \times 10^{-21} \text{ Farad}, \qquad [XV-4]$$

which is a 2:1 difference for the two methods of calculation.

These parameters are used to calculate the frequency of rotation of the electron in the hydrogen orbit. Using the first value of capacitance,

$$f_h = 1/[2\pi(LC)^{1/2}] = 1.017 \times 10^{18} \text{ Hz}. \qquad [XV-5]$$

The orbital velocity is determined by the radius of the electron orbit (5.29×10^{-11} m.), and the frequency of rotation is therefore found to be

$$v_h = \omega\, r = 3.38 \times 10^8 \text{ m/sec}. \qquad [XV-6]$$

Therefore, using the first value for capacitance for the calculation results in an electron orbital speed which is slightly <u>faster</u> than the speed of light.

Using the second value of capacitance,

$f_h = 7.19 \times 10^{17}$ Hz, [XV-7]

and we get

$v_h = 2.39 \times 10^8$ m/sec. [XV-8]

The second value of capacitance results in an electron velocity which is slightly <u>less</u> than the speed of light.

The variation of the speed for the two methods is within +13% of the speed of light for the first method and -25.5% for the second method. The exact value depends upon the capacitance between the electron and the proton. The capacitance between the proton and the electron is not clearly defined since they are not particles and their "size" is not well defined. If the forces of nearby atoms slow the electron, then the effect on the inductance of the electron orbit will produce a speed which is somewhat less than these values.

The capacitance varies inversely with the distance between the charges (the radius of the atom), while the inductance increases with the orbital path length. It was concluded earlier that the radius of the atom cannot increase significantly (without ionization taking place or the expansion of the mass). These factors provide clues as to the possible shapes of the electron orbits. It is also interesting to note that, by this method of calculation, the propagation time of free space [$(LC)^{-1/2}$] is proportional to the radius of the orbit and is therefore a <u>function of the speed of the orbital electron</u>, as had been reasoned earlier.

Such a high electron speed may seem surprising, since the slower orbital speed that was previously calculated using Newton's equation is much slower than the speed of light and has been the accepted standard. Using Newton's method of calculation, mass was regarded as non-electrical and constant, and inertial force and electrical force were equated. In the second method, Maxwell's equations for the electrical field were used without resorting to a calculation involving mass.

How else can the speed of the orbiting electron be determined? The ejection of orbital electrons does not always produce photons. When a voltage is applied across a non-conductor, the electron orbits are believed to distort into an ellipse. The elongation of the ellipse must reduce the speed of the electron. The slow application of voltage will therefore result in ionization and the ejection of an electron of lower velocity. On the other hand, if an electron is knocked from orbit with enough power, it may then eject at its maximum speed. But electrons do not move at the speed of light (no electron that moves at the speed of light has ever been detected). According to present theory, when the electron velocity reaches the speed of light, it can only exist as a photon, which is another clue to the mechanism by which photons are created. It is tempting to believe that the orbital speed of the electron is very close to the speed of light, since all of our observations point in that direction.

To thy speed add wings.

---Milton

CHAPTER XVI

Speeds Exceeding the Speed of Light

Einstein believed that <u>nothing</u> can travel any faster than the speed of light. His conclusions were, however, based primarily on energy considerations ($e = mc^2$), and his Theory of Relativity pictures space and time as being distorted by velocity and mass. The assertion has been made that any such apparent distortion is actually due to an illusory image of our surroundings, just as the earth can appear flat and as the universe seems to revolve about us. The Theory of Relativity "works" in the sense that the model fits the observation. But if the observation is the source of the problem, then other possibilities for the reality of the universe exist. In this case, if our measured data on outer space is corrupted by the measurement problem, then we must allow that the speed of light may not be as limited as is presently believed.

Einstein's theory is one of perception and perspective as applied to the geometry of space. He used visualizations to support his theory. Using his example of a rotating disk with one observer on the disk and one outside the disk, his conclusion was that the disk will appear to be distorted to the external observer due to the limited speed of light. He extended this argument to reach the conclusion that all of space is curved and, because energy is matter and matter is energy, there is <u>nothing</u> that can travel any faster than the speed of light. The Lorentz transformation is an equation that expresses the distortion of the dimensions of time and space with velocity, and he was thus able to express the amount of distortion explicitly using these equations. All of the elements of his theory fit together rather well and are quite believable.

Is it possible that Einstein's theory does not provide an accurate picture of the universe? For objects which are moving at very high speeds, is space distorted, or is it the image, itself, that is distorted? Consider the hypothetical scenario of a universe <u>without any optical radiation</u>, and wherein no object in the vicinity travels any faster than the speed of sound. If we did not have the ability to sense optical radiation and could only hear sounds emitted from moving objects, our view of the universe would be completely different. Our vision of space would condense, and our view of the universe would be limited to the earth itself, if we did not have the capability of leaving the earth's surface. Applying the same reasoning that was used for the Theory of Relativity, the spatial distortion would become much more severe, and even nearby moving objects would exhibit spatial distortion. If we could only sense the speed of objects using sound waves, and if any of the objects in our local vicinity did not travel any faster than the speed of sound, then perhaps we would reach the conclusion that <u>nothing could travel faster than the speed of sound</u>. Using a similar argument, and because electrical and optical radiation is the limiting factor, we can conclude that it is the speed of optical radiation that limits our vision of the universe. It is a *measurement problem*.

There is evidence to support this argument. It is known that time sampling of a moving object can present an illusory distortion. The moving picture camera is an example. Each photographic frame is a snapshot in time, and the replay of these pictures can be represented as a time distortion in slow or fast motion. Another example is the stroboscopic illumination, such as that of the timing light of the auto mechanic, which can make objects appear to rotate much slower than they actually move. The above scenarios are examples of the well-known sampling problem, which has been thoroughly investigated. The assertion can be made that the speed of the electrons of the atom that limits our time-sampling of the universe. The radiation of energy, upon which optical detection depends, is a function of the time that it takes for the electron of an atom to complete one or more orbital rotations. Thus each photon constitutes a time sample of radiation.

But if the speed of electrons is limited, and if all mass is constructed of atoms and molecules which contain these electrons, then how would it be possible for <u>anything</u> to travel any faster than the speed of light? First of all, if there is something there that we cannot sense accurately, then errors in measurements are produced. It is the limited speed of light that restricts the measurement of the higher speed of objects, and therefore our sensing method for high-speed objects is limited and inaccurate.

We saw that the Big Bang theory provides a plausible reason why it can be possible to exceed the speed of light. In Chapter IV, it was contended that all of the atoms in the universe are racing radially outwards along the bubble of the universe at speeds which are near the speed of light. This assertion is based on the present cosmological view that the universe is quite flat but in the form of a bubble. If evidence is some day obtained that corroborates the theory that the universe is flat but slightly curved, it will substantiate the bubble theory and lend credence to the proposition that the speed of light can be exceeded.

Using lasers, we can transmit light in very short bursts of energy; so short that they are sometimes called "photon torpedoes" since they appear in the shape of a torpedo when detected by optical instruments. If I shoot a photon torpedo to my left and simultaneously shoot another photon torpedo to my right, then how fast are they moving apart? Are they not moving apart at twice the speed of light? Ah, but you argue "If you were on one of these photon torpedoes, time and space would change such that the other photon torpedo would be moving away at exactly the speed of light?" Perhaps so, but I am not on either of these torpedoes, and, even if I was there, I would never see the other torpedo because the light from it would never catch up to me. While neither is traveling faster than the speed of light, they must be flying apart at twice the speed of light.

The Theory of Relativity dispenses with this problem by not allowing an absolute measurement. In the above argument, neither of the photon torpedoes were moving faster than the speed of light with respect to <u>my location in space</u>. So what if I am on a train and shoot a photon torpedo out the window? Is it not traveling slightly faster than the speed of light, both with respect to the my location on the train and a fixed point on the ground? No, the speed of light does not change with respect to either location, according to the Theory of Relativity, since time and space become distorted with movements that are near the speed of light. So the Theory of Relativity covers our human experiences, but it is rather confusing to try to visualize all of space being distorted by velocity and mass.

Ch. XVI - 186

We have claimed that the distortions of moving objects are produced by a measurement problem, due to the limited speed of the electron in its orbit about the nucleus of the atom. Let us consider another assertion. According to present theory, energy radiation does not occur unless there is transition in the orbit of an electron. This process takes a finite interval of time since the orbiting electron is traveling at a given speed. It can be inferred that it is the <u>movement of the electron with respect to the proton</u> (as it changes its orbit) that establishes the speed at which the optical wave moves through space. Therefore, the wave will propagate at the speed of light as determined by the distance traveled by the electron, and the time taken to move that distance. If the circular velocity of a rotating field is unlimited, as with Concept #2 of Chapter VIII, then the radiation of light is directly related to the instantaneous movements of the orbital electron. If Concept #2 holds up under further investigation, then a new theory of radiation may result.

Then what about group motion of electrons? It was argued that the group motion of electrons, in an antenna, can also produce radiation. How does that relate to the above method of generating light? There is no conflict, since the antenna can only produce waves of a greater wavelength than the size of the orbit of an atom. The length of an antenna is much greater than the radius of an atom, and both the distance and the time to traverse a cycle of radiation increases. Antennas simply radiate in a somewhat different manner.

A type of measurement distortion does occur, however, for any movement that occurs between the radiating atom and an external detector. An optical detector, moving toward the atom, will traverse a cycle of radiation in less time than when it is moving away from the atom. Thus the wavelength, the distance traveled by the wave for a full cycle, changes with movement between the radiating source and the detector. The result is a form of distortion of the signal. The change in wavelength is the now familiar "Doppler effect." Therefore, the process of moving either the source of radiation or the detector can produce a measurement problem.

So is it possible for radiation to ever exceed the <u>relative</u> speed of light? Not for electrons moving around protons at their present speed. If we could get them to move faster in their orbits, it is conceivable that they would radiate energy at speeds greater than the speed of light. However, the dynamics of the atom, as we know it today, would not permit a faster electron speed since the electron could not be contained by its attraction to the proton. Such a task would be an enormous undertaking, even if it were possible. The Big Bang, if there was one, may have set the electrons in motion about the nucleus of atoms, at a fixed speed which was determined by the force of the explosion. Another possibility is that the orbital velocity was somehow established before the initial event of the Big Bang occurred.

Assuming that it is possible to exceed the speed of light, would it be possible to measure such speeds accurately, if at all? Light travels so fast that we humans cannot judge the magnitude of the speed directly with our senses. Any measurement, that is used to calculate the speed of travel of an optical event, depends upon a hypothetical theory and special equipment. We do not have a fixed vantage point in space at which we can measure the speed of expansion of the bubble of the universe. This fact, along with the arguments in the above paragraphs, imply that anything moving faster than the speed of light must be located at very remote distances in order to obtain a measurement.

The Doppler frequencies of optical radiation from stars in outer space provide velocity information on their movements. We know that space is expanding, and at a very fast rate. The sun is moving at 155 miles per second within our own galaxy, the Milky Way, which in turn is estimated to be traveling through its location in the universe at a rate of 370 miles per second. The furthest galaxy is believed to be traveling away from us at a velocity of about one-eighth the speed of light. If the universe proves to be closed, but nearly flat, then the universe is curved, and the bubble theory assumes new importance. In that case, the speed at which the universe is expanding, in a radial direction, is likely to be much faster than the presently measured speeds, and very likely exceeding the speed of light

With the many objects moving so fast in so many directions, our observations of time and space are distorted to some degree. A fast moving object traveling at right angles to an observer is distorted in space, while objects moving directly towards or away from us are distorted in time. The geometry of the universe must therefore be somewhat different from that which we observe. Perspective is essential to our view of the universe. For instance, from our position at the earth's surface, the earth appears somewhat flat, but the astronaut far above the earth's surface sees it quite differently. The earth did not change, but the vision of it did.

While the Theory of Relativity presents one picture of reality, it is based on the distortion of space by gravitational forces and velocities. As we learn more about gravity and radiation, physical theories will inevitably change. Our vision of the universe depends on the extent and accuracy with which we can sense or measure the universe. The goal is to develop an approach that will present a vision of reality that fits the facts, and these facts depend on sensing or measurement. If there is something that is happening of which we are ignorant, due to the fact that we cannot sense it, then our vision of the universe will be incomplete. Inaccurate measurements can also change our models. Today, the measurements of astronomers and the analysis of cosmologists is changing our vision of the universe as this book is being written, which is due in large part to the pictures of outer space that are being sent to us from the Hubble space telescope. The universe is in the process of being mapped to a much higher degree of accuracy. With this new data, the Big Bang and expanding bubble theories will be confirmed or denied. If it is confirmed, then we should consider the possibility that the speed of light can be exceeded.

The hydrogen engine is the workhorse of the universe. Hydrogen is the lightest, fastest, most active atom of all, and radiation measurements indicate that the universe is pretty well filled with it. It is also the most simple atom with but one electron and one proton which is quite fortunate since it is therefore much easier to analyze. The analyses of the prior chapters, based primarily on the hydrogen atom, tend to support the conclusion reached here: *it is the speed of these electrons within atoms that controls the speed of light and all radiation.* We sense fast movements by this radiation, and our perspective depends upon the limited degree of sampling that light provides. By recognizing that our model of the universe is in error due to a measurement problem, we can make the necessary corrections and form a new vision of the universe. The Theory of Relativity is not refuted, as a model, but it does not provide the clearest vision of the events occurring around us. If events are actually occurring around us at a speed faster than the speed of light, would it be possible for us to sense them? Perhaps not.

The whirligig of time.

---Shakespeare

CHAPTER XVII

Is the Electron a Wave?

One of the confusing properties of electrons that has baffled physicists is that an electron sometime behaves as a particle and sometimes as a wave. In most cases, the electron acts as a particle. Electrons can be ejected from excited atoms, and one method of inducing electrons to leave an atom is to heat a material. When a substance is heated, electrons leave the surface. Some materials emit more electrons than others at a given temperature. The electron gun within a cathode ray tube, such as the one inside our TV set, works in this manner, and the accelerated electrons that are produced by appear to act as a beam of <u>particles</u>. However, when electrons pass through a thin piece of matter or a thin gap between two pieces of matter, they exhibit properties similar to that of a <u>wave</u>. The present view of the electron must therefore accommodate two separate situations, since this dual-property problem has not yet been solved.

The basis of a unified field theory was developed in the previous chapters. A major theme of this book is that <u>everything</u> in the universe consists of electrical fields. Even voids in space have weak fields and waves moving through them. Any moving field can be considered to be a wave, and so a moving electron with its associated field must therefore be a wave. The wave theory is supported by measured evidence that shows that the strength of the electric field around the electron does not have a discrete shape, diminishing gradually with distance, and a moving field is a wave.

The fields, established between electrons and other protons and electrons, extend out an unmeasurable distance. Therefore, the field about an electron depends upon every other charge in the universe and all of their locations, and the center of this field (or wave) is simply the location of the electron. The nearest charges exert by far the greatest effect. The proton is similar to the electron, having a field that decreases in intensity with distance, but of opposite polarity. While the proton is highly attracted to the nucleus of the atom, and is tightly bound within it, the electron has less attraction to the nucleus and has greater freedom of movement. The center of the field associated with an electron or proton is simply a point, and the near field of this point can, however, be thought of as a particle.

The intensity of the field is inversely proportional to the square of the distance to other charges, so the nearby charges control the majority of forces on the electron. However, considering the immense amount of charge in the universe, even distant charges can have a significant effect if they exist in sufficient numbers. Hence every field and every wave depends upon everything else that exists in the universe.

An optical wave, such as a beam of light, experiences diffraction when it passes through an aperture, spreading out in waves of several significant wavelengths and producing patterns which can be observed. Similarly, an electron (and its field) experiences "diffraction" when it passes through a thin non-conducting crystal with its many tightly-packed electrons and protons. If the crystal were not thin, the electron might not make it through the crystal since its energy would be absorbed by the molecules that it contacts during its passage. The exiting electron beam spreads out into a ring of waves having a *diffraction pattern* (mathematically, a sinx/x pattern) similar to that of an optical wave. The similarity between the actions of the wave and the electron have led to the belief that the electron can also act as a wave. The diffraction process for the electron is well known, but the causes that produce the effects are not well understood. The following analysis presented is one version of how electron diffraction can is produced:

As an electron passes through a crystal, it is attracted repelled by other electrons and attracted to protons in the nuclei of nearby atoms. When it approaches too close to the nucleus of an atom, it encounters a negative force. Therefore the moving electron experiences forces of attraction and repulsion as it passes through an atom. The electron must therefore vibrate during its trip through the crystal.

The atoms or molecules of the crystal are evenly spaced in a lattice structure. The atom is very similar to an electronic oscillator, since they both exhibit harmonic behavior. Electronic analysis and measurement shows that oscillators easily lock together with just a minor amount of coupling. Oscillators that are oscillating at different frequencies will lock together in a common frequency if their frequencies are close enough together. If the electrons in these atoms are locked together in movement, then the electron will experience a *periodic force* that increases in level as it moves through the crystal. It is therefore reasonable to infer that the moving electron will therefore experience an oscillating movement as it passes through the crystal, and at certain angles these oscillations will reach a maximum variation, while at certain other angles the amplitude of the oscillations will be a minimum.

A narrow beam of electrons, which are moving in a straight line are deflected in this manner during passage through the crystal, and, therefore, they form diffraction patterns by virtue of the harmonic forces exerted upon them during their trip through the crystal. This result does *not*, by itself, infer that an electron is a wavelike field, but only that the effect of the forces within the crystal act on the electron in a harmonic manner. Electron beam diffraction patterns appear similar to those of x-ray photons, which <u>are</u> waves. Experiments of this type have led to the belief that an electron can sometimes act as a wave by virtue of they generate similar diffraction patterns.

There is another experiment that illustrates the wavelike characteristics of an electron. When a moving electron passes through a <u>narrow slit</u> in a piece of metal it also generates a diffraction pattern. In order for the individual atoms to exert a harmonic force on the moving electron, as above, the slit must be sufficiently narrow. Light, which is accepted as consisting of waves, creates a similar effect. The similarity in the two patterns further illustrates the wavelike characteristics of electrons.

The indications are that the electron is a wave, but of what type of wave? The field about an electron depends on nearby protons and other electrons. And since the electron behaves in a manner similar to that of X-rays, it is possible that the electron itself may be a wave oscillating at a very high frequency. The electron possesses a small amount of inertia, and since rotational fields are associated with inertia, the electron can thus be pictured as a rotating wave. Analysis, using the methods of quantum physics, indicates that the electron has the property of spin. The electron has two directions of spin, which is an indication that the field may be planar, since turning the electron around reverses the direction of spin.

But the electric field, as we know it, also has fixed properties. Static electricity and lightning can be produced from a static field of unchanging level or polarity. Is each little electron actually a tiny rotating field, behaving similar to the action of a gyroscope, stable in space until moved by an external force, or is it simply a particle? We still do not know very much about the electron, since we have no definitive method of sensing what is happening within the near field of the electron.

The above analysis presents a dilemma. The evidence leads us to believe it plausible that a single electron is, itself, a wave rotating in space. And yet the electrical field, as we know it, exists only between electrical charges. If the electron is a rotating wave, then we may be able to sense its movement by the oscillating electrical force that it exerts on other charges, and yet vibrating forces of this type are not know to exist. If the electron is a wave, then we may need to redefine the electrical field, since our present view of the electrical field is based on two opposite charges rather than a single charge. The proton is even more complex in its characteristics, and the evidence suggests that fractional charges exist within it (quarks).

Again, each answer leads to new questions. While there is evidence to suggest that the electron is a wave, we still do not know very much about the details of its construction. Measurements of atoms, especially the little electrons and protons, provide insufficient information to obtain a detailed picture of what is happening in the interior . The interior of an electron is much smaller and much more difficult to probe, by any means known. No electron has ever been imaged. We are once more at the edge of knowledge, which is another illustration that we still have much more to learn about electricity.

Man's capacities have never been measured.

---Thoreau

CHAPTER XVIII

The Electricity Within Us

Each moving electron, rotating around the nucleus of an atom, can be considered as a current that is internal to the atom. While it is not the same type of electrical current as that which flows within a wire, the definition of current as a moving charge can also be applied to the atom. It occurred to me that it would be an interesting exercise to estimate the electrical current moving within our bodies, as based on the equivalent current within atoms.

The current flowing within the hydrogen atom depends upon the charge of the electron, which is well known, its radius, and the frequency of rotation. The frequency of rotation is estimated using the electrical constants, L and C, which are calculated for the spatial geometry of the atom. The frequency of rotation, as calculated in Chapter XV is 1.017×10^{18} Hz. The resulting current flowing in the hydrogen atom is found to be 1.0237 Ampere. This value of current will be used as the basis for estimating the total electrical current flowing within the atoms of our bodies.

The majority of the molecules within our body are of water, which is H_2O, containing two hydrogen atoms and one oxygen atom. The oxygen atom has eight electrons circulating around its nucleus, and, for the purposes of this calculation, the speed of rotation of these electrons will be assumed to be assumed to be the same as that of the hydrogen atom. Therefore the each water molecule has 10.237 Amperes of current flowing within it.

From chemistry, we know that the mass of one mol of H_2O weighs exactly ten grams. Each mole of any compound contains 6.022×10^{23} molecules, so each gram of water contains 6.1647×10^{24}

Amperes of current. A person weighing 150 pounds has 68,040 grams of matter within his/her body. Therefore, the electrical current flowing within the atoms of matter of a 150 pound person is approximately 4.19×10^{29} Amperes, which is some 400 billion, billion, billion Amperes! We are all human powerhouses, having even greater power within the nuclei of our atoms. Too bad we cannot tap some of a small of that energy which is not being used. However, almost all of this electrical power is securely locked up within atoms that do not want to let loose of their electrons.

    Heavier persons have a proportionately greater electrical current contained with their atoms and vice versa. It is amusing to think that the mass of one's body can be measured by the current flowing within one's body. A billion amperes is one Giga-Ampere, so a billion, billion, billion Amperes can be written as one G^3A, and a 150 pound person would have 419 G^3A of current flowing within them. A 200 pound person would then have 559 G^3A, etc. Anyone who has ever claimed to be a dynamo is underestimating themselves.

CHAPTER XIX

Putting it All Together

A new view of the universe has emerged from the information presented in the previous chapters. A universe that consists solely of electrical fields circulating throughout, with each atom affected in some way by every other atom in every other moon, planet, star or galaxy. A circular universe. A harmonic universe. A universe that is surprisingly empty, but with forces so strong that they exercise the imagination to its fullest.

And yet ours is a universe which is also extremely placid, much more so than it would seem. Human senses are highly limited, clouding our vision, and making it extremely difficult to determine exactly what is happening around us. Our surroundings appear mostly solid and unmoving to the touch, and nearby moving objects travel at very slow speeds. It is difficult for us to imagine a universe composed of unseen electrical fields; one that is so empty and yet so powerful, dynamic, harmonious and cohesive.

Up to now, no one has yet been able to establish an all-encompassing, unified field theory that accounts for the various forces that exist in nature. Scientists have examined the "strong forces" within the nucleus of atoms in an attempt to show that they produce the very weak force of gravity. Strong forces can oppose one another, and some success has been made in getting the addition right, such that the forces add together by the right amount to produce the gravitational force. Just how this process is supposed to occur is not clear. Subatomic particles, *gravitons*, are believed to be the source of this reaction. However, not a single graviton has ever been detected, even though many investigations have been conducted in search of them.

The main stumbling block, in the attempts on the part of scientists to solve the secret of gravity, has been the general acceptance of an important assumption. It has been previously assumed that the electrical charges within the atom cancel outside of the atom. Since there is an electron for every proton in the atom, and because the neutrons are electrically neutral, no external electrical forces are supposed to exist. If these forces cancel, then they must survive the test for cancellation. A test of this type indicates that they do not cancel. In Chapter VII, it was proposed that a universal force exists in the universe and that it is *electrical*. An explanation was provided as to how the gravitational force is produced by the electrical forces within the atom. As a result, we no longer need the elusive graviton to provide the force of gravity.

A method of fitting the gravitational force into the unified theory was believed to be the last piece of the puzzle. The now famous Grand Unified Theory (GUT) is not yet complete because of this reason. More recently, scientists have been concentrating on the effects of subatomic particles, and there is still no accepted <u>complete</u> unified field theory. If gravity does not depend on subatomic particles, then unification follows directly from the concept of a universal force that is electrical. It is possible that subatomic particles are also a form of the electrical field.

Einstein came very close to reaching the conclusion that there is a universal field and that it is electrical. He outlined the many similarities between the gravitational force and the electrical force. Einstein noted that Maxwell's equations of electromagnetic radiation fit the Theory of Relativity and stated, "These laws fit the frame of the special relativity theory, since they are invariant with respect to the Lorentz transformation." Theodor Kaluza, in 1919, rewrote Einstein's equations for the gravitational force and noted that Maxwell's equations were present in the solution. Others have also observed the similarities and tried to combine the two forces, but none of these theories have yet been successful in providing sufficient proof that they are valid.

Energy is matter and matter is energy according to relativity theory. But light is energy, and it is electrical energy, so why not conclude that mass is electrical? The problem of relating *matter and field* in a sensible concept was, however, a problem without a solution. There was no way to resolve the enormous difference between the almost imperceptible gravitational force and the enormous electrical force within the atom. And so what happens as we reach a point close to the atom? Where does mass end and the electrical field begin? "Could we not slightly modify our equations so that they would be valid everywhere, even in regions where energy is enormously concentrated?" (Einstein). The results of Chapter IX provide an answer to that very difficult question. The proof that the gravitational force is a consequence of the electrical forces within the atom is given in Appendix II.

This means that everything in the universe is electrical in nature. We can now look at mass in a completely different light. If this new gravitational theory can be accepted, then the previously stated dilemmas are resolved and one must conclude that *mass is a characteristic of a moving electrical force field within the atom*. The new mathematical expression, representing mass, now includes two second-order terms involving the radii of the electron and proton orbits. Based on this result, the inference can be made that mass is proportional to the area of the orbit (or volume of the atom). In other words, mass is related to the size of the atom and the number of charges within it. This is a sensible and meaningful result.

But what about the nucleus of the atom? Is it, too, electrical? The assertion that it is electrical was made earlier. The protons and neutrons in the nucleus are electrical, even though the neutron has no electrical charge. It was shown that a hydrogen atom, compressed into a smaller size, can exert the same external forces as that of a normal hydrogen atom if the rotation of the proton within the neutron rotates in a circle of the proper diameter. Therefore, it is possible that the nucleus of an atom is similar to a compressed hydrogen atom, in which case the charges within the nucleus of an atom are also electrical.

Then what about the various subatomic particles? Are they also electrical? According to current theory, fractional electrical charges, in the form of quarks, are present within the proton. While this has never been proven, assume it to be the case. Three quarks, of positive and negative charge within the proton, form three dipoles (a tripole?), and if they rotate, then the analytical methods of Appendix II can be used to determine the resulting external forces of the proton. But what about the many other subatomic particles, such as the neutrino, baryons, mesons, etc. We suggested that the neutrino may simply be a balanced rotating dipole which has no external forces. As to the others, well, school is still out on them (most of them do not seem to be around very long, anyway). They could simply be smaller, mostly unstable rotating fields.

Matter is not simply a group of particles but a collection of tiny sources of electrical power, so many in number that the total power in even a handful of dirt is enormous. All forces tend to seek balance, and these forces are all electrical. Negative and positive are in rotation in each atom, and the electrical fields tend to cancel outside the atom, although not totally. The force between two atoms increases greatly as they come together, which can also account for the so-called molecular force which holds solids together. As the distance between atoms increases, the force between them decreases, and, when the distance is large, the force is at the level of the gravitational force..

Gravity is a consequence of the action of the weak internal electrical force of the atom, producing an even weaker external force. Comparing gravitational force with electrical force results in a difference which is so great that it does not appear to be able to ascertain that the two are related. The weak electrical force is about 10^{39} times stronger than the gravitational force. This amount of difference is hard to imagine and difficult to comprehend. The diameter of an atom is 10^{-10} meter, while the universe may be some 4×10^{27} meters in diameter. The forces of the universe are so delicately balanced that the ratio of forces is similar to the ratio of sizes if we were to balance the entire width of the universe on a knife edge only one atom thick!

Using another analogy to demonstrate the extent of this ratio, we will substitute time as the variable, since it can be measured much more accurately than other parameters, such as force, voltage, or current. If, for instance, an instrument can measure time to within one picosecond (a millionth of a millionth of a second), it would take some 3 billion years to achieve a ratio of the two time periods that equates to the above number. The forces of the universe are obviously incredibly well balanced and comparatively placid (lucky for us).

The theories that have been proposed in this book depend upon the thing that is most easily measured: *force*. The main proposition of this book is that there is only one force, and that is electrical force. Every action is a consequence of the dynamic forces within atoms. In order to be able to produce a new vision of radiation and matter, it was necessary to examine the underlying method by which radiation occurs. It was shown that atomic orbital electrons are probably moving at speeds close to the speed of light. The time that it takes for an atom to radiate energy depends on the amount of time that elapses as an orbital electron changes its path.

We can only sense what is happening around us by the radiation that passes our way. The smallest particle of radiation is the photon, and each photon acts to "sample" the world about us. If the sampling rate is too slow, then the accuracy of measurement degrades and the image becomes distorted. Therefore, parametric distortions are a result of the limited speed of the electrical fields that exist in the universe rather than being produced or controlled by gravity. In other words, the apparent warping of time and space is produced by a measurement problem.

Measurement errors are commonly caused by the "sampling problem." Engineers are quite familiar with the term, since problems of this type occur in digital communications, radar, telephony, and instrumentation. Not everyone, however, is knowledgeable about sampling problems, and examples were provided in earlier chapters. Mechanics, for instance, time the ignition system of a car by observing the rotating flywheel of a car. The flywheel shaft appears smooth when it rotates fast, due to the slow response time of the human eye. The motion of the shaft can be observed by viewing it with illumination by a stroboscope. The flashing light of the stroboscope samples the flywheel at a much faster speed than the human eye can detect directly, and, as the sampling speed approaches that of the flywheel, it appears to be rotating slowly. When the exactly correct sampling speed is obtained, the shaft appears to stop its motion. The apparent speed of the flywheel depends on the sampling speed, thereby distorting our vision of the dynamics of the action. A similar phenomena occur for the photons of light from radiation, resulting in distorted illusions of a fast-moving object.

On earth, trains, automobiles, and planes move at what we consider high speed, but these velocities are negligible compared to those of the planetary movements, rotating galaxies, cosmic explosions, or the rate of inflation of the universe produced by the Big Bang. But even the gravitational forces of the Big Bang are quite docile compared to the power contained _within_ atoms. If just the nuclei in the atoms of your own body were to fly apart, it would destroy the earth. A dynamic reaction of this type may have happened at the first instant of the Big Bang. Luckily, what we consider as chaos, turmoil, tempest, disorder, disruption, tumult, and upheaval are very tame in comparison with the power and order that exists in the universe. The power that we possess is insignificant to the other energies that are flying around us. Objectively speaking, we are infinitesimal dots in time, space, dimension, or energy.

Other contemporary theorists have varying beliefs. John Casti concludes that humans are quite special, a "flicker of intelligence" in the universe. David Bohm, a protégée of Einstein's, believes that everything that we experience is but a ghostly image beyond space and time. Michael Talbot envisions a holographic universe. What we believe depends on what we can accept as the truth. There is some degree of evidence for all of these diversionary theories.

Chapter IV dealt with the Big Bang theory, which has gained acceptance over the past several years. With the many theories of the origin of the universe, there are various conceptual versions of the Big Bang. I chose the *expanding bubble* theory as the most likely possibility, since it has the greatest amount of evidence to support it. The Doppler shift of energy from distant stars has provided valuable information for analyzing the dynamics of the universe. Hubble was the first to measure the Doppler shift of the galaxies and the important information that he obtained indicates that the universe is expanding.

Hubble used the Doppler shift to calculate the rate of separation of distant galaxies. From this data, he estimated the age of the universe to be 2×10^9 years. Various contributors have increased this estimate of age by nearly ten times. Calculations, based on the estimated amount of matter in the universe, indicate that the universe is flat, and the flatness could be attributed to galactic matter forming the surface of a sphere, in which case the universe is considered to be "closed." For a closed universe, if the curvature of the universe is <u>assumed</u> to be .01 degree, the calculations indicated a diameter of the universe of 4×10^{27} meters, an outward (radial) velocity of 2.4×10^8 m/s (0.8 times the speed of light), and an age of the universe of $.3 \times 10^{11}$ years (about ten times older than current estimates). Is all of this really to be believed? With a grain of salt. The Hubble space telescope will produce more accurate data to map the universe, enabling a more accurate estimate of curvature.

The Big Bang theory, as analyzed above, exercises Einstein's proposition that nothing can travel faster than the speed of light. Even though the speed of atomic electrons is at or near the speed of light, all matter may be <u>speeding radially outward along the bubble at or near the velocity of the atomic electrons</u>.

But how can such high velocities be possible? Again, the effect of any measurement must always be considered since we cannot "sample" space at any faster a rate than is permitted by the rotation of the fastest electron in the atoms of the materials that we use for our instrumentation. Photons appear to have length, and they travel at a limited speed. The *sampling theorem* of electronics states that, in order to sample a repetitive (or harmonic) signal and preserve the information in the process, the sampling rate must be at least twice the frequency of the signal. An object, or the radiation from an object, moving faster than the speed of light would have to be sampled at two distant points along its traverse which are fixed in the frame of reference of the space that we are measuring. The sampling time is the transit time of the photons from the source to the detector. At least two samples must be obtained during the interval of measurement, and many more samples are necessary to provide additional information as to the shape of the object. The length of the photon is also important, since it must illuminate only a fraction of the object during the measurement time. It is easy to see how objects can appear distorted when they move at sufficiently high velocities.

Since we are traveling at a point (on our spot on the bubble) which is not fixed with respect to the center of the universe, it is not possible, from this perspective, to obtain an <u>absolute</u> measurement of the velocity of expansion of the universe, so it is not just a measurement problem, but also a spatial location problem. We may never know, for sure, whether or not we are all flying through space at speeds which are at, or above, the speed of light. As with most cosmological theories, we have little proof, one way or the other.

If the premise that it is possible to exceed the speed of light is accepted, then how is the model of the universe that currently exists affected? If we accept this possibility, it removes certain difficulties, and it also provides some much needed explanations. It certainly supports the idea of the Big Bang, the bubble theory, and a universe that is expanding at a radial speed near the speed of light. It is also much easier to visualize than a distorted space/time relationship. The main drawback? That the speed of light can be exceeded is in conflict with Einstein's Theory of Relativity, a contradiction that most will probably find difficult to accept.

The bubble theory depends on the Theory of Gravity. But is Newton's gravity equation correct or complete? The expansion of the galaxies indicate a potential problem, due to the rotary velocities of the stars, their mass, and the inertial force. The analysis of Chapter XII also shows that the Newton's theory may be incomplete, or even inaccurate. If his theory is not exact, then many cosmological theories may fall apart, like a row of dominos. This possibility is not highly likely, but it is still to be considered. Until the difficulties are resolved, the Big Bang theory, with the universe structured as a bubble, seems to be the most believable cosmological theory of them all.

Other problems with basic theories of physics exist. In Chapter VIII, the rotating field of the electron/proton dipole of the hydrogen atom was analyzed with enigmatic results. It was shown that the rotational speed of the field can exceed the speed of light (in a circular or tangential direction), unless the field changes shape with rotational velocity. If the rotating field of the dipole does not exceed the speed of light as the radius of the field increases, the field must distort in some way such that the speed is limited to the speed of light In order for the speed of the field to always conform to the Theory of Relativity, the shape of the field becomes unbelievably distorted. On the other hand, if the speed of light can be exceeded (for this particular geometry), then the other enigmas are removed, and the idea of a measurement problem is substantiated. Such a possibility does not necessarily conflict with the Theory of Relativity, since it applies to radiation, and the rotating dipole does not radiate, except under certain conditions. The characteristics of the external field of a dipole, rotating at high speed, therefore assume great importance. No far field test data on fast-rotating dipoles could be found in the literature. We must await further developments in science to resolve the dilemma.

Electrically, the hydrogen atom is similar to an electronic oscillator, and it was suggested that the electrons within adjacent atoms would tend to lock together in frequency, as do electronic oscillators. Synchronous motion is necessary, for dissimilar atoms that share electrons, in order for the electrons to not interfere with one another. This factor has significance in tests which involve bombardment of materials by beams of electrical particles and X-rays, due to the energy patterns produce by the exiting rays. Do all of the atoms in a material lock together to form a symphony of harmonic motion? Such an effect would have profound consequences, and, if confirmed, would contribute to our further understanding of the characteristics of matter. Again, we do not yet have all of the evidence to make the determination.

And then there is the inertial force. Einstein, you will remember, equated the gravitational force with the inertial force. He reasoned that the force of gravity is opposed to the inertial force, and therefore whatever it is that produces one must also produce the other. This reasoning does not necessarily follow. The analysis indicates that the inertial force, like the gravitational force, is also a function of the electron orbit, since the two forces are closely related The contention was made that the inertial force is a <u>consequence</u> of the gravitational force. Cause always comes before effect. When a gravitational force acts on an atom, the electron is affected first, since it is rotating about the proton(s). It therefore changes orbit slightly, and the orbital path becomes distorted (squashed, so to speak). The distortion of the orbit changes its gravitational attraction correspondingly, and the effect is to oppose the gravitational force. The gravitational force acts as the cause, and the inertial force is the effect.

But what about the inertia of the proton and the electron? The difficulties in analyzing the electron were elucidated in the previous chapter. Physicists have a little more data concerning the proton. The proton is also a "charged particle," and the four-body system representing two hydrogen atoms, that was analyzed in Appendix II, becomes a three-body system due to the three charges (quarks within the proton) rather than the four charges existing in two hydrogen atoms. It may be possible to also apply these methods to the analysis of the proton. Before this can be done, however, the characteristics of the so-called quarks must first be determined, if they indeed do exist. Perhaps some day they will find one. The inertial characteristics of basic charge, the electron and the proton, are subject to further study, which is also beyond the effort presented here.

The current belief is that the electron of the atom is traveling at a fraction of the speed of light (10^6 m/s using Bohr's early version of quantum physics or 2.23×10^6 m/s using Coulomb's law and Newton's equation). In Chapter XVI, Maxwell's equations were employed in calculating the velocity of the electron in the hydrogen orbit. Two possible values for the electron velocity were obtained: 3.39×10^8 m/s and 2.39×10^8 m/s using two methods to calculate the value of the electrostatic field. <u>This is a variation of an order of magnitude from the present accepted speed</u>. On the other hand, if Maxwell's equations do not apply within the atom, then a new important enigma will surface, since a new set of field equations will be required for the fields within the atom.

Assuming Maxwell's equations do apply within the atom, the result of the analysis is appealing, since the speed of a fast orbital electron orbit can approach the speed of light, and it is more logical that a photon can be generated by a high velocity electron rather than a slow one. Electrons appear to act as waves, in certain ways, and, when they are accelerated to speeds approaching the speed of light, they exhibit all of the characteristic of a wave. The ejection of an electron from its orbit generates a photon which is a wave (a form of a field), and so, if the orbital electron is traveling at the speed of light and is ejected, it would presumably be a wave --- a photon.

Another enigma that occurs in the application of quantum theory, is that the electron is pictured with its orbit passing through the nucleus of an atom. But the nucleus is also believed to have an energy barrier, while an barrier prevents the possibility of such an event. The atom is a nonlinear system, and nonlinear systems of this type can be analyzed using the Mathieu's equation of nonlinear analysis. The proper application of Mathieu's equation to electron orbitals provides a mathematical description of a system, wherein the electron does not fall into the nucleus of the atom. The forces on the electron must be such that repulsion exists as it nears the proton, while attraction takes place as the orbital radius increases. Nonlinear analysis can be highly complex, resulting in equations that defy analysis, even with a computer of high computing power. Such an analysis is too complex to be included in a book of this type, but the use of the Mathieu equation is another example of how classical analysis can be applied to the problems of atomic physics.

If we are to understand gravity, and if the gravitational force is an electrical force, then it would serve is well to have a good understanding of electricity. Maxwell's equations relate the magnetic force to the electrical force. But these equations depend on the definition of magnetism. In Chapter X, the equations for the magnetic field were re-written in terms of the velocity of moving charges rather than the rate-of-change of electrical charge. The magnetic terms were thus eliminated, showing that the electrical field is the only field that need be considered.

Electrical conduction, even after many years of investigation, is not fully understood. An analysis of the location of elements in the Periodic Table of the Elements that conduct electricity was presented. The best conductors are found in unfilled s^1 energy shell. The assertion was made that the atom must not be bound too tightly to its neighboring atoms in order to be a good conductor; that the conducting atom must be as likely to give up an electron equally as well as it accepts one

And then there is the question of how slow moving charges, such as in an antenna, can produce an electrical field that expands at the speed of light. The assertion was made that this phenomena is due to the *group motion* of the electrical charges and they move within the antenna.

In order to fully understand the dynamics of the universe, the process of radiation is of importance. One of the failings of classical theory has been the inability to explain the spectral radiation of materials or predict the spectral patterns. Atoms have characteristic spectral patterns that conform to the wavelength of radiation, and these patterns do not appear to correlate to other types of manmade radiation. Quantum theory, on the other hand, has been able to provide calculated spectral patterns that match radiation fairly well.

Hydrogen, our most simple element, has only a single proton and a single electron, and yet it emits numerous discrete spectral lines. In quantum theory, these wavelengths correlate with multiples of the atomic orbit, and the equations are based on energy transfer. The shapes of the atomic orbits can be determined by using Fourier analysis, which was mentioned earlier in the discussion of our "circular universe." Quantum theory has, however, produced some contradictions in the analysis of radiation. The nucleus of the atom is said to have an energy wall through which the rotating electron cannot penetrate, while the orbital patterns for radiation show the electron passing through the center of the nucleus of the atom.

In chapter XIII, it was shown that classical analysis could provide spectral patterns similar to that of radiation spectra when parametric modulation was applied. Frequency modulation is a type of parametric modulation that is used in FM radio transmitters. Parametric modulation of this type fits the conditions that exist within the atom since it is equivalent to an electric field modulating an electron moving in its orbit in the atom.

Parametric modulation generally produces *symmetrical* spectral envelopes, however, while atomic spectra are seldom symmetric in shape. The assertion was made earlier that the lack of symmetry was produced by a nonlinear effect. If the orbitals are analyze by applying Mathieu's equation, the resulting spectral envelopes tend to bend toward a higher or lower frequency. With the proper amount of bending, spectral envelopes of the type observed can be obtained. The discrete spectral patterns of radiation produced by atoms is accounted for in this manner.

One of the goals was to establish theories that are not so abstract, and Bohr's idea of an atomic radius of any size do not fit this proposition. The desired result is obtained by simply substituting the *orbital path length* for the radius. The spacing of the spectral radiation lines indicate harmonic motion, which means that the electron path must be repetitive in some manner.

The electron can move in various paths in three dimensions, which presents numerous possibilities for achieving harmonic repetition. Orbital path shapes are therefore quite important, and there is more than one way to fulfill this requirement and still preserve the essence of Bohr's argument. The electric field depends on the electron orbit, and the field represents energy. Moving fields can radiate energy (waves), and the radiation of energy is basic to the dynamics of the universe. Einstein showed that mass represents energy, and therefore *the force of gravity is related to the electron orbit*; a conclusion reached earlier by a somewhat different reasoning.

Physicists have recorded portions of the radiation spectrum of energy emitted from atoms. This data is then analyzed mathematically

Ch. XIX - 209

to determine the atomic orbitals. It is in this manner that orbitals passing through the center of the atom was calculated. Bohr contended that energy radiation occurs as a result of a change in the radius of the electron orbit. A more likely possibility is that radiation occurs when there is a *change in the path length* of the electron orbit. The atomic spectral data can then be used to relate the *change in the orbital shape* to the electron orbit of the atom. The electron revolves in a fixed orbit and radiates only when the orbit changes to a degree such that the electron does not have to pass through the nucleus of the atom. This assumption resolves the above dilemma,

The characteristics of radiation are a function of the spatial parameters of the orbit. There are three dimensions to space, and the *direction of the radiation* is important. The *shape of the orbital path* also affects the spectral characteristics. Many inferences as to atomic forces and interactions can be drawn from the shape of the electron orbitals and the forces that produce radiation. Various orbital paths and radiation characteristics were illustrated in Chapter XIV. Different orbital paths sometimes exhibit spectral characteristics that resemble one another. The importance of the shape of atomic orbits is not to be underestimated.

The analysis of chapter XI indicates that there is yet another type of force in the universe that has not yet been observed; at least not directly. As the electron moves in its orbit around the nucleus, it creates a vibrational force at a very high frequency. This force is much stronger than the average force of gravity between two atoms. The frequency of this vibration is dependent on the rotational speed of the electron and its orbital path. Since orbital speeds are very fast, high vibrational force fields that proliferate in the universe, producing atomic vibration at frequencies too high to detect (in the range of 10^{18} Hz). These forces are vectorial (directional), and point in various directions. Are these forces vital to the thermal agitation of atoms and molecules? If so, it will then be necessary to show why the amount of vibration decreases with temperature. It may be possible that the electron orbit changes shape with temperature. For now, we can only contemplate the possibilities and look to the future for answers.

And so, everything in the universe consists of tiny electrical fields, most of which are in motion. They tend to aggregate, creating masses, and the attractive force of gravity and their inertial forces are proportional to the square (or perhaps the cube) of the radius of the electron orbit of the atoms. Radiation depends upon the orbital path, and a planar orbit has different radiation characteristics than a spherical path. When the electron changes radius, energy is either absorbed or radiated. For radiation to occur, the electron must briefly attain the speed of light, and laboratory tests of moving electrons show evidence that electrons do exhibit this property.

The exact speed of the electron, in its orbit around the nucleus of the atom, is not known, but it is presently believed to be a fraction of the speed of light. The calculations of chapter XV, using Maxwell's equations, produced a speed much closer to the speed of light. For a calculated rotational frequency of 10^{18} Hz, each moving atomic electron constitutes a current of one Ampere of electricity. The human body is mostly water, so someone weighing 150 lbs. contains a total atomic current of 4.19×10^{29} Amperes! The electrical fields of the "weak force" of atoms within our body are powerful indeed. The power within the nuclei of these atoms are far stronger yet, and so the total electrical power that is swirling within us defies the imagination.

Rather than believing that space and time are truly altered by high speeds and large masses, it appears much more likely that these apparent distortions are an apparition, formed as a result of the limitations of our ability to sense what is happening. The limited speed of orbital electrons within atoms and the fixed speed of light, while being extremely fast, limit any measurement method available to us. By realizing that the apparent distortions of space are produced by the limits of measurement that are available to us, and through the determination of the exact properties of radiation, it may be possible to straighten out these aberrations, and obtain a clearer vision of the universe.

So, is the Theory of Relativity invalid? In any case, the speed of light still has a maximum value, due to the limited speed of orbital electrons. However, if you are willing to accept the argument that light provides a limited accuracy of measurement of the space around us, then we must conclude that a measurement problem exists. The Lorentz equations still apply to moving objects for all measurements, but rather than showing how space, itself, becomes distorted, they are showing how the picture of space is deformed. A distortion of this type is similar to that obtained by looking into a curved mirror. The assertion is that the measurement problem produces the *apparent distortion* of time and space. Space is not truly distorted; it only appears distorted for objects moving at high speeds, such as galaxies and the high speed gases within them. The dimensions themselves do not actually change. And, if space is expanding outward at a sufficiently high rate, then earth, itself, may be traveling faster than the speed of light.

The scientific approach requires evidence to corroborate a hypothesis. What is meant by "corroborate?" It means that either no contradictions exist, or else the contradictions are acceptable. In the case of quantum physics, certain contradictions are present, and yet the theories are accepted, since meaningful results are obtained. The theories put forth within this book will have to survive similar tests. For inner space, tests can be performed. A mathematical proof that the gravitational force can be produced by the electrical forces within the atom is presented in the appendix. This is an important method of testing, and, if no flaws are to be found in the analysis, then the theory may gain acceptance. Other hypotheses were presented, in some case with minimal proof. The tests, however, require extensive resources, and possible future efforts are discussed in the next chapter. For outer space, such evidence will be very difficult to obtain for the reasons given earlier. A map of the universe may provide some answers. If the universe proves to be in the form of a bubble, and if the mapping measurements and calculations indicate that the bubble is expanding at, or greater than, the speed of light, then we can begin to think of a new, absolute universe.

Chapter XX

AFTERWARDS

And so that is the end of the story. So where do we go from here? Do you believe that the secret of gravity now been solved? The mathematical proof of Appendix II indicates that the external electrical forces of the hydrogen atom are more than sufficient to account for the gravitational force. It was necessary to modify Rutherford's model of the atom, by allowing the proton to rotate in a rather large orbit, in order to equate the electrical force to the gravitational force. If the equations have been set up properly, then the secret of gravity has been solved, and the impact on science may be enormous.

You are invited to run these simulation programs on your own computer and check the results. The "epsilon" or "delta"(the smallest number allowed in the floating point calculation) of the mathematical simulation program must be very low in order to reduce the simulation errors. A floating point epsilon of one part in one billion, billion, billion, billion, billion, billion, billion was selected, and yet the calculations of Appendix III produced final errors which limited the exact determination of the radius of the proton orbit. A more powerful computer may also be required to produce results which are more accurate and have reasonable run times (the simulations took about two hours to run on a Pentium at 120 MHz). Even if mathematicians prefer to solve the equations directly, the results must be verified by computer simulations so that any assumptions that are made do not result in errors.

It was shown that the orbital path of the rotating electron of the atom has many consequences. For the Rutherford model, the axial alignment produced a negative force between two atoms. It may be that the attraction, or repulsion, between certain molecules can be attributed to the orbital alignment of atoms. For instance, certain plastics tend to cling to other materials. It may be that some attractive forces, normally attributed to static electricity, may be produced by distorted orbits or orbital alignments.

The analysis of Chapter XII indicates the anomalies of the

gravitational forces within the earth. Any modification of, or addition to, Newton's theory of gravitation will have an impact on cosmological theories, which can have far-reaching implications. It is possible to test some of the hypotheses that have been presented. The Cavendish experiment should be repeated, and the forces within the two masses metered at various points within the balls as they approach one another (perhaps with a bag of metal balls). Masses of different shape should also be tested, since it was argued that the center of gravity is not what it has been presumed to be. If the definition of the center of gravity can be extended to remove the existing enigmas and dilemmas, then Newton's Theory of Gravity can be modified, by addition, to take into consideration the variation in gravitational forces throughout the mass. If the gravitational constant changes at great distance, then we may have found the reason why galaxies are not flying apart as the present calculations would indicate.

In Chapter VIII, the field wave of a non-radiating hydrogen atom was analyzed, and some important enigmas surfaced. It is possible to perform a different type of test to see if the speed of a wave can exceed the speed of light. The characteristics of the field of a dipole, rotating at high speeds, can be measured at various distances. If the field wave rotates at a speed faster than the speed of light (in a tangential direction along a radius from the center of rotation), then Einstein's theory must be modified to allow the speed of light to allow the speed limit to be modified for rotational dipoles. If the speed does reach a limit, then the field intensity curves change with the speed of rotation; a result that seems unreasonable. Either of these possibilities are of importance, and the results of such test may result in new models of the atom and the universe.

The length of a photon is of particular importance, since it determines the minimum sampling time of any event that can take place. The number of wavelengths in a single photon provides a clue as to the path of the electron as it changes orbit when it creates a photon.

Not all of the problems that were considered in this book were solved. Actually, more new problems were revealed than were solved. It is likely that some of them will not be solved in my lifetime, and perhaps in yours, too. In particular, the mystery of the electron and proton may never be uncovered, since the tools for measurement of

Ch. XX - 215

these tiny electrical "things" may not be available to us (never say never). We need a tool that is much smaller than either of them to do the job. Subatomic particles?----Mmmm, perhaps.

Some new ideas, some new conclusions, some new assertions, and a somewhat different view of the universe.

Something to think about

APPENDIX I

The Calculation of the Gravitational Force Between Two Hydrogen Atoms

Newton's gravitational force equation is given by

$$f_g = G\, m_1 m_2 / d^2 \qquad [\text{I.A-1}]$$

where

$$G = 6.67259 \times 10^{-11}\ \text{nt-m}^2/\text{kg}^2 = \text{gravitational constant} \qquad [\text{I.A-2}]$$

$$m_1 = m_2 = m = 1.6735 \times 10^{-27}\ \text{kg} = \text{mass of the hydrogen atom.} \qquad [\text{I.A-3}]$$

Therefore,

$$K_g = m^2 G = 1.8687 \times 10^{-64}\ \text{nt-m}^2 \qquad [\text{I.A-4}]$$

and

$$f_g = K_g / d^2 = \text{gravitational force between two hydrogen atoms.} \qquad [\text{I.A-5}]$$

The force of gravitational attraction between the two hydrogen atoms as a function of their separation is plotted in Figure IX-1. **The gravitational attraction between the two atoms is 1.8687×10^{-62} newton at a separation of 0.1 meter.** This amount of separation is used for comparison with the electrical forces that are calculated in Appendix II.

APPENDIX-II

CALCULATION OF THE ELECTRICAL FORCE BETWEEN TWO HYDROGEN ATOMS

II.A The Electrical Force Equation

The electrical force equation for two point charges is given by:

$$f_e = q_1 q_2 / 4\pi\varepsilon_0 d^2 = K_e / d^2 \qquad [\text{II.A-1}]$$

where

$$q = 1.6 \times 10^{-19} \text{ Coulomb} \qquad [\text{II.A-2}]$$

$$\varepsilon_0 = 8.85 \times 10^{-12} \text{ farad/meter.} \qquad [\text{II.A-3}]$$

Re-arranging terms,

$$K_e = q^2 / 4\pi\varepsilon_0 = \text{electrical force constant,} \qquad [\text{II.A-4}]$$

$$K_e = 2.3019 \times 10^{-28}. \qquad [\text{II.A-5}]$$

The above equations describe the force between two separated electrical charges.

II.B The Force Between Two Hydrogen Atoms Whose Electron Orbits are Aligned in the Same Plane

The Rutherford model of the hydrogen atom pictures the electron rotating about the nucleus, which contains a single proton. The force between two hydrogen atoms depends upon the alignment of the two electron orbits. The first configuration to be considered has the two orbits aligned in the same plane as shown in Figure IX-2 in Chapter IX.

It is very important to set up the equations for any mathematical model very carefully. The methods used to develop equations is particularly important in this case since a very tiny error function is hidden in the electrical force equation for multiple electrical charges. Equation [II.A-1] contains only two terms, and one of these is a constant. Therefore, the electrical force is a function of the distance between the charges and is a nonlinear inverse square function. We are seeking the error function of the subtraction of charges external to the atoms, and this function will depend upon how the electrical field varies with distance from the center of the atom.

The protons of the two hydrogen atoms are assumed to be fixed in space, with the electron of each atom rotating about the protons and the orbits co-planar. Like charge repel, and unlike charges attract. Therefore, four forces control the overall force between the two atoms, two attract, and two repel. The summation of these four forces is the error function that we are seeking. Since the electrons are rotating, three of the forces vary with time, and an average must be obtained. The average is obtained by integrating the force equation over one or more cycles of rotation. Two cycles have been chosen in order to show that the function is repetitive.

It was found that computation errors develop rapidly as the distance between the atoms increase. Therefore, the radius of the atom is exaggerated to a value which lessens the computational error, and a comparatively short distance between atoms is used. It must be scaled down to the actual value by assuming that the error varies as inverse square law. Thus a small amount of error in the calculations can be expected.

The following parameters have been selected:

$r_h = 0.0001$ meter = exaggerated orbital [II.B-6]
radius of the hydrogen atom

$d = 0.1$ meter = selected distance between atoms. [II.B-7]

Since these computations were compiled on a personal computer (PC), the number of points for computation is also limited,

$k = 0..2000$ = number of points for the numerical calculation. [II.B-8]

For two cycles of rotation,

$\theta(k) = \pi(k/500)$ = incremental angle of rotation of the electron in the first atom. [II.B-9]

The electron of the second atom is not assumed to rotate at the same frequency, and the frequency is chosen to be a harmonic so that averaging does not induce unwanted errors. The distance between the various charges of the two hydrogen atoms are

$d_{pp}(k) = d = 0.1$ meter = distance between protons [II.B-10]

$d_{ee}(k) = [r \bullet e^{j \bullet (\pi \bullet k/500)} + d + r \bullet e^{j \bullet (\pi \bullet k/250)}]$

= distance between electrons, [II.B-11]

$d_{ep1}(k) = [r \bullet e^{j \bullet (\pi \bullet k/500)} + d]$ [II.B-12]

= distance between the electron in the first atom and the proton in the second atom,

$d_{ep2}(k) = [d + r \bullet e^{j \bullet (\pi \bullet k/250)}]$ [II.B-13]

= distance between the proton in the first atom and the electron in the second atom.

The above equations are written in complex vector notation, and, to calculate the force between the two atoms, only the real part of the equation is used. Therefore,

$$f_{hh}(k) = Ke[\ 1/Re^2(d_{ep1}(k)) + 1/Re^2(d_{ep2}(k)) - 1/Re^2(d_{pp}(k)) - 1/Re^2(d_{ee}(k))] \qquad [II.B-14]$$

= instantaneous force between the two hydrogen atoms.

The effect of the distance between the atoms is illustrated in Figure IX-3 with the electrical constant, Ke, deleted from equation [II.B-14].

The cyclical nature of the force is obvious as seen from the above graph. In order to obtain the average force between the two atoms, the force equation is integrated over two full cycles,

$$F_{hh}(k) = \frac{1}{2000} \int f_{hh}(k)dk \qquad [II.B-15]$$

The results are plotted in Figure IX-4. It would appear that the force between the two hydrogen atoms sum to zero. This is not the case, however,

$$F_{hh}(2000) = 2.99251872 \cdot 10^{-7} \qquad [II.B-16]$$

= scaled average force between the two hydrogen atoms.

To determine the actual force between the atoms, we must include the electrical constant and re-scale for the proper orbital radius,

$$F_{hh}(act) = K_e \cdot (r_h/r)^2 \cdot F_{hh}(2000) \qquad [II.B\text{-}17]$$

$$= 1.9349754 \cdot 10^{-47} \text{ newtons.}$$

= total average force between the two hydrogen atoms with their orbits aligned in the same plane.

Note that not only do the electrical charges not cancel, they are much larger than the gravitational force as calculated in Appendix I.

II.C The Force Between Two Hydrogen Atoms With Axial Alignment

The second configuration to be considered has the two orbits aligned axially as shown in Figure IX-5 in Chapter IX. The procedure is identical to that of the previous section, using the same constants. Two cycles have been chosen in order to show that the function is repetitive.

The distance between the various charges of the two hydrogen atoms are

$$d_{pp}(k) = d = 0.1 \text{ meter} = \text{distance between protons}$$
$$[II.C\text{-}18]$$

$$d_{ee}(k) = [r \cdot e^{j \cdot (\pi \cdot k/500)} + d \cdot e^{i \cdot (\pi/2)} + r \cdot e^{j \cdot (\pi \cdot k/250)}]$$

$$= \text{distance between electrons,} \qquad [II.C\text{-}19]$$

$$d_{ep1}(k) = [r \cdot e^{j \cdot (\pi \cdot k/500)} + d \cdot e^{i \cdot (\pi/2)}] \qquad [II.C\text{-}20]$$

= distance between the electron in the first atom and the proton in the second atom,

$$d_{ep2}(k) = [d \cdot e^{j \cdot (\pi/2)} + r \cdot e^{j \cdot (\pi \cdot k/250)}] \quad [\text{II.C-21}]$$

= distance between the proton in the first atom and the electron in the second atom.

The above equations are written in complex vector notation, and, to calculate the force between the two atoms, only the real part of the equation is used. Therefore,

$$f_{hh}(k) = Ke[\, 1/Re^2(d_{ep1}(k)) + 1/Re^2(d_{ep2}(k)) - 1/Re^2(d_{pp}(k)) - 1/Re^2(d_{ee}(k))\,] \quad [\text{II.C-22}]$$

= instantaneous force between the two hydrogen atoms.

The effect of the distance between the atoms is illustrated in Figure IX-6 with the electrical constant, Ke, deleted from equation [II.C-22]. The shape of this curve differs from that of Figure IX-3.

Again, the cyclical nature of the force is clear as seen from the above graph. In order to obtain the average force between the two atoms, the force equation is integrated over two full cycles,

$$F_{hh}(k) = \frac{1}{2000} \int f_{hh}(k) dk \quad [\text{II.C-23}]$$

The results are plotted in Figure IX-7. It again appears that the force between the two hydrogen atoms sum to zero, and, again this is not the case.

IIA - 224

$$F_{hh}(2000) = -3.99993 \cdot 10^{-10} \qquad [\text{II.C-24}]$$

= scaled average force between the two hydrogen atoms.

To determine the actual force between the atoms, we must include the electrical constant and re-scale for the proper orbital radius,

$$F_{hh}(act) = K_e \cdot (r_h/r)^2 \cdot F_{hh}(2000) \qquad [\text{II.C-25}]$$

$$= -2.586374 \cdot 10^{-50} \text{ newtons.}$$

= total average force between the two hydrogen atoms with their orbits aligned axially.

For axial alignment, the force is one of repulsion. Notice that the force is much less than that of equation [II.B-17].

II.D   The Force Between Two Hydrogen Atoms With Orthogonal Alignment

The last configuration to be considered has the two orbits aligned orthogonally as shown in Figure IX-8 in Chapter IX. The method is the same as before.

The distance between the various charges of the two hydrogen atoms are

$$d_{pp}(k) = d = 0.1 \text{ meter} = \text{distance between protons} \qquad [\text{II.D-26}]$$

$$d_{ee}(k) = [r \cdot e^{j \cdot (\pi \cdot k/500)} + d + r \cdot e^{j \cdot (\pi \cdot k/250)}]$$

= distance between electrons, [II.D-27]

$$d_{ep1}(k) = [r \cdot e^{j \cdot (\pi \cdot k/500)} + d] \qquad \text{[II.D-28]}$$

= distance between the electron in the first atom and the proton in the second atom,

$$d_{ep2}(k) = [d + r \cdot e^{j \cdot (\pi \cdot k/250)}] \qquad \text{[II.D-29]}$$

= distance between the proton in the first atom and the electron in the second atom.

The above equations are written in complex vector notation, and, to calculate the force between the two atoms, only the real part of the equation is used. Therefore,

$$f_{hh}(k) = Ke[\ 1/Re^2(d_{ep1}(k)) + 1/Re^2(d_{ep2}(k)) -$$

$$1/Re^2(d_{pp}(k)) - 1/Re^2(d_{ee}(k))] \qquad \text{[II.D-30]}$$

= instantaneous force between the two hydrogen atoms.

The effect of the distance between the atoms is illustrated in Figure IX-9 with the electrical constant, Ke, deleted from equation [II.D-30]. The shape of this curve differs from that of Figure IX-3 and IX-6.

Again, the cyclical nature of the force is clear as seen from the above graph. In order to obtain the average force between the two atoms, the force equation is integrated over two full cycles,

$$F_{hh}(k) = \frac{1}{2000} \int f_{hh}(k) dk \qquad \text{[II.D-31]}$$

The results are plotted in Figure IX-10. It again appears that the force between the two hydrogen atoms sum to zero, and, again this is not the case.

$$F_{hh}(2000) = 1.99601 \cdot 10^{-7} \quad [\text{II.D-32}]$$

= scaled average force between the two hydrogen atoms.

To determine the actual force between the atoms, we must include the electrical constant and re-scale for the proper orbital radius,

$$F_{hh}(act) = K_e \cdot (r_h/r)^2 \cdot F_{hh}(2000) \quad [\text{II.D-33}]$$

$$= 1.2906286 \cdot 10^{-47} \text{ newtons.}$$

= total average force between the two hydrogen atoms with their orbits aligned orthogonally.

The force between the two atoms is attractive, as was the case for the planar alignment.

II.E AN ESTIMATE OF THE AVERAGE FORCE BETWEEN TWO HYDROGEN ATOMS FOR THE THREE ORBITAL ALIGNMENTS

The alignment of all of the hydrogen atoms is not known, so an assumption is required in order to make an estimate. For a random alignment of the orbits, all of the possible angles of rotation in three planes must be considered. We will settle for an easier approach and a rougher estimate by simply averaging the three forces. From the previous sections.

$$F_{planar} = 1.9349754 \times 10^{-47} \text{ newton} \quad [\text{II.B-17}]$$

$$F_{axial} = -2.5863743 \times 10^{-50} \text{ newton} \quad [\text{II.C-25}]$$

$$F_{orthogonal} = 1.2906286 \times 10^{-47} \text{ newton} \qquad [\text{II.D-33}]$$

and averaging,

$$\mathbf{F_{electrical} = 1.0743392 \times 10^{-47} \text{ newton.}} \qquad [\text{II.E-34}]$$

The gravitational force was determined as

$$\mathbf{F_{gravitation} = 1.8687 \times 10^{-62} \text{ newton.}} \qquad [\text{II.E-35}]$$

Not only do the electrical forces outside the atom not cancel, they are much higher than the calculated gravitational force.

APPENDIX-III

ALTERING THE RUTHERFORD MODEL TO REDUCE THE ELECTRICAL FORCE TO CONFORM TO THE GRAVITATIONAL FORCE

III.A The Electrical Force Equation

The Rutherford model of the atom is to be modified by allowing the proton of the hydrogen atom to rotate so that the external electrical force of the atom are reduced to a level compatible with the gravitational force, The basic charge parameters of Appendix II do not change.

$$f_e = q_1 q_2 / 4\pi\varepsilon_0 d^2 = K_e / d^2 \qquad [\text{III.A-1}]$$

where

$$q = 1.6 \cdot 10^{-19} \text{ Coulomb} \qquad [\text{III.A-2}]$$

$$\varepsilon_0 = 8.85 \cdot 10^{-12} \text{ farad/meter.} \qquad [\text{III.A-3}]$$

Re-arranging terms,

$$K_e = q^2 / 4\pi\varepsilon_0 = \text{electrical force constant,} \qquad [\text{III.A-4}]$$

$$K_e = 2.3019 \cdot 10^{-28}. \qquad [\text{III.A-5}]$$

The above equations describe the force between two separated electrical charges.

III.B The Force Between Two Hydrogen Atoms Whose Electron Orbits are Aligned in the Same Plane for the Modified Rutherford Model

The Rutherford model of the hydrogen atom pictures the electron rotating about the nucleus, which contains a single proton. The proton will now be allowed to rotate. The computer simulation program is not sufficiently powerful to determine the exact orbital radius of the proton, so the radius of the proton is increased until the summed electrical force is within the error limit of the program. As before, the first configuration to be considered has the two orbits aligned in the same plane as shown in Figure IX-2 in Chapter IX.

As was the case for the example of Appendix II, the radius of the atom is exaggerated to a value which lessens the computational error, and a comparatively short distance between atoms is used. It must be scaled down to the actual value by assuming that the error varies as inverse square law.

The following parameters have been selected:

r_e = 0.0001 meter = exaggerated orbital radius of the electron of the atom [III.B-6]

r_p = 0.7428802•r_e = orbital radius of the proton of the atom [III.B-7]

d = 0.1 meter = mean distance between atoms. [III.B-8]

$k = 0..2000$ = number of points for the numerical calculation. [III.B-9]

For two cycles of rotation,

$\theta(k) = \pi(k/500)$ = incremental angle of rotation of the electron in the first atom. [III.B-10]

The electron of the second atom is not assumed to rotate at the same frequency, and the frequency is chosen to be a harmonic so that averaging does not induce unwanted errors. The distance between the various charges of the two hydrogen atoms are

$$d_{pp}(k) = [d + r_p \cdot e^{j \cdot (\pi \cdot k/250)} + r_p \cdot e^{j \cdot (\pi \cdot k/500)}] \quad \text{[III.B-11]}$$
= distance between protons

$$d_{ee}(k) = [r_e \cdot e^{j \cdot (\pi \cdot k/500)} + d + r_e \cdot e^{j \cdot (\pi \cdot k/250)}] \quad \text{[III.B-12]}$$
= distance between electrons,

$$d_{ep1}(k) = [r_e \cdot e^{j \cdot (\pi \cdot k/500)} + d + r_p \cdot e^{j \cdot (\pi \cdot k/250)}] \quad \text{[III.B-13]}$$
distance between the electron in the first atom and the proton in the second atom,

$$d_{ep2}(k) = [r_e \cdot e^{j \cdot (\pi \cdot k/250)} + d + r_p \cdot e^{j \cdot (\pi \cdot k/500)}] \quad \text{[III.B-14]}$$
= distance between the proton in the first atom and the electron in the second atom.

The above equations are written in complex vector notation, and, to calculate the force between the two atoms, the absolute distance between atoms is used. Therefore,

$$f_{hh}(k) = Ke\{[\,1/(abs(d_{ep1}(k)))^2 + 1/(abs(d_{ep2}(k)))^2]$$

$$[1/(abs(d_{pp}(k)))^2 - 1/(abs(d_{ee}(k)))^2]\} \quad \text{[III.B-15]}$$

= instantaneous scaled force between the two hydrogen atoms.

The instantaneous force is integrated as before,

$$F_{hh}(k) = \frac{1}{2000} \int f_{hh}(k)dk \qquad [\text{III.B-16}]$$

and

$$F_{hh}(2000) = 2.2886265 \cdot 10^{-8} \qquad [\text{III.B-17}]$$

= scaled average force between the two hydrogen atoms.

To determine the actual force between the atoms, we must include the electrical constant and re-scale for the proper orbital radius,

$$F_a = K_e \cdot (r_h/r)^2 \cdot F_{hh}(2000) \qquad [\text{III.B-18}]$$

= - 1.479835683348·10⁻⁴⁸ newtons.

= total average force between the two hydrogen atoms having rotating protons and with their orbits aligned axially.

Note that the force is now repulsive.

III.C The Force Between Two Hydrogen Atoms With Axial Alignment

The second configuration to be considered has the two orbits aligned axially as shown in Figure IX-5 in Chapter IX. The procedure is identical to that of the previous section, using the same constants.

The distance between the various charges of the two hydrogen atoms are

$$d_{pp}(k) = [r_p \bullet e^{j \bullet (\pi \bullet k/500)} + d \bullet e^{j \bullet (\pi/2)} + r_p \bullet e^{j \bullet (\pi \bullet k/250)}]$$
$$= \text{distance between protons} \quad \text{[III.C-19]}$$

$$d_{ee}(k) = [r_e \bullet e^{j \bullet (\pi \bullet k/500)} + d \bullet e^{j \bullet (\pi/2)} + r_e \bullet e^{j \bullet (\pi \bullet k/250)}]$$
$$= \text{distance between electrons,} \quad \text{[III.C-20]}$$

$$d_{ep1}(k) = [r_e \bullet e^{j \bullet (\pi \bullet k/500)} + d \bullet e^{j \bullet (\pi/2)} + r_p \bullet e^{j \bullet (\pi \bullet k/500)}] \quad \text{[III.C-21]}$$

= distance between the electron in the first atom and the proton in the second atom,

$$d_{ep2}(k) = [r_e \bullet e^{j \bullet (\pi \bullet k/250)} + d \bullet e^{j \bullet (\pi/2)} + r_p \bullet e^{j \bullet (\pi \bullet k/500)}] \quad \text{[III.C-22]}$$

= distance between the proton in the first atom and the electron in the second atom.

The above equations are written in complex vector notation, and, to calculate the force between the two atoms, only the absolute value of the distance iis used. Therefore,

$$f_{hh}(k) = Ke\{[\, 1/(abs(d_{ep1}(k))^2) + 1/(abs(d_{ep2}(k)))^2] -$$
$$[1/(abs(d_{pp}(k)))^2 - 1/(abs(d_{ee}(k)))^2]\} \quad \text{[III.C-23]}$$

= instantaneous force between the two hydrogen atoms.

In order to obtain the average force between the two atoms, the force equation is integrated over two full cycles,

$$F_{hh}(k) = \frac{1}{2000}\int f_{hh}(k)dk \qquad \text{[III.C-24]}$$

It again appears that the force between the two hydrogen atoms sum to zero, and, again this is not the case.

$$F_{hh}(2000) = -8.0327 \cdot 10^{-11} \qquad \text{[III.C-25]}$$

= scaled average force between the two hydrogen atoms.

To determine the actual force between the atoms, we must include the electrical constant and re-scale for the proper orbital radius,

$$F_b = K_e \cdot (r_h/r)^2 \cdot F_{hh}(2000) \qquad \text{[III.C-26]}$$

$$= -5.193994233311053 \cdot 10^{-51} \text{ newtons.}$$

= total average force between the two hydrogen atoms with their orbits aligned axially.

For axial alignment, the force is one of repulsion.

III.D The Force Between Two Hydrogen Atoms With Orthogonal Alignment

The last configuration to be considered has the two orbits aligned orthogonally as shown in Figure IX-8 in Chapter IX. The method is the same as before.

The distance between the various charges of the two hydrogen atoms are

$$d_{pp}(k) = [r_p \bullet e^{j \bullet (\pi \bullet k/500)} + d + r_p \bullet e^{j \bullet (\pi \bullet k/250)}] \quad \text{[III.D-27]}$$
= distance between protons

$$d_{ee}(k) = [r_e \bullet e^{j \bullet (\pi \bullet k/500)} + d + r_e \bullet e^{j \bullet (\pi \bullet k/250)}] \quad \text{[III.D-28]}$$
= distance between electrons,

$$d_{ep1}(k) = [r_e \bullet e^{j \bullet (\pi \bullet k/500)} + d + r_p \bullet e^{j \bullet (\pi \bullet k/250)}] \quad \text{[III.D-29]}$$
= distance between the electron in the first atom and the proton in the second atom,

$$d_{ep2}(k) = [r \bullet e^{j \bullet (\pi \bullet k/250)} + d + r_p \bullet e^{j \bullet (\pi \bullet k/500)}] \quad \text{[III.D-30]}$$
= distance between the proton in the first atom and the electron in the second atom.

The above equations are written in complex vector notation, and, to calculate the force between the two atoms, the absolute value of distance is used. Therefore,

$$f_{hh}(k) = Ke\{[\, 1/((d_{ep1}(k)))^2 + 1/(abs(d_{ep2}(k)))^2 -$$

$$[1/(abs(d_{pp}(k)))^2 - 1/(abs(d_{ee}(k)))^2]\} \quad \text{[III.D-31]}$$

= instantaneous scaled force between the two hydrogen atoms.

In order to obtain the average force between the two atoms, the force equation is integrated over two full cycles,

$$F_{hh}(k) = \frac{1}{2000} \int f_{hh}(k) dk \quad \text{[III.D-32]}$$

It again appears that the force between the two hydrogen atoms sum to zero, and, again this is not the case.

$$F_{hh}(2000) = 2.22964469 \cdot 10^{-8} \qquad [\text{III.D-33}]$$
= scaled average force between the two hydrogen atoms.

To determine the actual force between the atoms, we must include the electrical constant and re-scale for the proper orbital radius,

$$F_{hh}(act) = K_e \cdot (r_h/r)^2 \cdot F_{hh}(2000) \qquad [\text{III.D-34}]$$

$$= 1.4892373319735 \cdot 10^{-48} \text{ newtons.}$$

= total average force between the two hydrogen atoms with their orbits aligned orthogonally.

The force between the two atoms is attractive.

III.E AN ESTIMATE OF THE AVERAGE FORCE BETWEEN TWO HYDROGEN ATOMS FOR THE THREE ORBITAL ALIGNMENTS

We estimate the force between the atoms by simply averaging the three forces. From the previous sections.

$$F_{planar} = -1.479835683348 \cdot 10^{-48} \text{ newton} \qquad [\text{III.B-18}]$$

$$F_{axial} = -5.193994233311053 \cdot 10^{-51} \text{ newton} \qquad [\text{III.C-25}]$$

$$F_{orthogonal} = \ = 1.4892373319735 \cdot 10^{-48} \text{ newton} \qquad [\text{III.D-34}]$$

and averaging,

$$F_{electrical} = -1.3704261744987 \times 10^{-52} \text{ newton.} \qquad [\text{III.E-34}]$$

= external electrical force between the two atoms.

The above force that is calculated turns out to be <u>repulsive</u>, but the calculated value is so low that it is probably in the numerical noise of the program. The lowering of the force is proven.

APPENDIX IV

MAXWELL'S EQUATIONS

Maxwell's four equations are given by:

$$\nabla \cdot \mathbf{D} = \rho \qquad \text{[IV-1]}$$
$$\nabla \cdot \mathbf{B} = 0 \qquad \text{[IV-2]}$$
$$\nabla \times \mathbf{E} = -\partial \mathbf{B}/\partial t \qquad \text{[IV-3]}$$
$$\nabla \times \mathbf{H} = \mathbf{J} + \partial \mathbf{D}/\partial t \qquad \text{[IV-4]}$$

where

$\mathbf{D}$ = electric displacement $\quad \mathbf{E}$ = electric field strength
ρ = charge density $\quad \mathbf{H}$ = magnetic field strength
$\mathbf{B}$ = magnetic field density $\quad \mathbf{J}$ = current density.

The wave equation in a medium with zero conductivity and

ϵ = permittivity
μ = permeability

can be derived from the above equations,

$$\nabla^2 = \mu\epsilon \, \partial^2 \mathbf{E}/\partial t^2, \qquad \text{[IV-5]}$$

and

$c = \sqrt{\mu\epsilon}$ = the velocity of the wave

$= 3 \cdot 10^8$ m/s in a vacuum.

APPENDIX V

THE LORENTZ TRANSFORMATION

According to the Theory of Relativity, there is no absolute velocity of anything moving through space. The velocity of light is the same in any reference system, and nothing can travel any faster. As a consequence, the dimensions of an object must vary with velocity. Applying the Lorentz Transformation,

$$x = x_0 \sqrt{1 - v^2/c^2} \qquad \text{[V-1]}$$

The length of an object, of initial length is x_0, varies with the velocity at which it is moving, $v(t)$. The length of the line is reduced along its line of motion as the velocity increases (with respect to a point of measurement). Newton's laws hold up when the Lorentz transformation and the Theory of Relativity are applied to moving objects.

VA - 238

BIBLIOGRAPHY

John Gribbin, *In Search of the Big Bang*, New York, Bantam Books, 1986
(Contains a wealth of historical information on cosmology)

Donald Goldsmith and Nathan Cohen, *Mysteries of the Milky Way*, Chicago, Contemporary Books, 1991
(A well-written description of the universe and easy reading, it includes the stages of nuclear fusion in a massive star)

Nigel Henbest, *The Exploding Universe*, New York, Macmillan Publishing Co., Inc., 1979
(Presents excellent illustrations and photographs of space, and it is also easy reading)

James S. Trefil, *The Moment of Creation*, New York, Charles Scribner's Sons, 1983
(Traces the history of the universe from the first instant of the Big Bang)

Albert Einstein and Leopold Infeld, *The Evolution of Physics*, New York, Simon and Schuster, 1938
(Probably Einstein's most simple writing, a book that all scientists should read)

John L. Casti, *Paradigms Lost*, New York, Avon Books, 1990
(Gives details of the contributions of many scientists to the various models of the universe with various visions of reality)

Michael Talbot, *The Holographic Universe*, New York, Harper Collins Publishers, 1992
(An interesting portrait of a universe that consists mostly of holograms)

Charles Oatley, *Electric and Magnetic Fields*, London, Cambridge University Press, 1976
(Concise analysis of electrical fields and the application of Maxwell's equations)

W. J. Cunningham, *Introduction to Nonlinear Analysis*, New York, McGraw-Hill Book Company, Inc. 1958
(The best book ever written on nonlinear analysis)

Einar Hille, *Analytic Function Theory*, New York, Blaisdell Publishing Company, 1959
(Highly mathematical technique for analyzing complex functions)

Ruel V. Churchill, *Complex Variables and Applications*, New York McGraw-Hill Book Company, 1960
(The standard text on complex variables)

Harald Friedrich, *Theoretical Atomic Physics*, New York, Springer-Verlag, 1990
(Illustrates the complexity of current quantum theory)

Athanasios Papoulis, *The Fourier Integral and Its Applications*, New York, McGraw-Hill Book Company, Inc., 1962
(A great text on the Fourier integral)

Amnon Yariv, *Quantum Electronics*, New York, John Wiley and Sons, 1989
(Another master scientist with practical experience)

John E. Gibson, *Nonlinear Automatic Control*, New York, McGraw-Hill Book Company, Inc., 1963
(A "must" book for anyone involved in nonlinear analysis)

David H. von Seggern, *CRC Handbook of Mathematical Curves and Surfaces*, Boca Raton, CRC Press, Inc., 1990
(This book illustrates hundreds of mathematical curves)

Murray F. Gardner, *Transients in Linear Systems*, New York, John Wiley & Sons, Inc., 1956
(A classic book that will persevere forever)

David Halliday and Robert Resnick, *Physics for Students of Science and Engineering*, New York, John Wiley and Sons, Inc., 1960
(These authors do an excellent job of presenting basic physics in a logical manner and with explicit examples)

Rogers D. Rusk, *Introduction to Atomic and Nuclear Physics*, New York Appleton-Century-Crofts, Inc., 1958
(A good basic book that provides a general description of the dynamics within an atom)

Physical Science Study Committee, *Physics*, Boston, D. C. Heath and Company, 1960
(Lots of good stuff on physics with some historical information)

Linus Pauling, *College Chemistry*, San Francisco, W. H. Freeman and Company, 1950
(From the master of the molecule, it also contains certain physical descriptions not to be found elsewhere)

Henry F. Holtzclaw, Jr., William R. Robinson, and Jerome D. Odom, *General Chemistry with Qualitative Analysis*, Lexington, MA, 1991
(A good, basic presentation of the electron structure within atoms)

242

INDEX

Aher 54
Ampere 22, 195, 196, 210
Aristotle 16, 20, 29, 36
atomic orbitals 10, 77, 126, 163, 209
atomic spectra 10, 155, 162, 176, 208
Big Bang .... 9, 16, 50-Ch. IV - 54, 61, 63-Ch. IV - 65, 66, 145, 185, 187, 188, 201-Ch. XIX - 204, 241
black body 26, 27, 156
black hole 36, 125, 146
Bohr 27, 32, 34, 64, 69, 71, 76, 77, 108, 143, 157, 158, 175, 177, 209
bubble of the universe 185, 187
bubble theory .... 51, 57, 58, 64, 65, 185, 187, 202-Ch. XIX - 204
capacitance .... 98, 143, 159, 179-Ch. XV - 181
Casti 35, 202, 241
Cavendish 147, 214
classical analysis 35, 69, 95, 124, 155, 157, 158, 161, 165, 178, 206, 208
close packing 130
Copernicus 21, 28, 29
Coulomb's Law 114, 116, 137, 179, 206
Curies 23
De Broglie 77
Democritus 19
deuterium 70, 75
dielectric constant 98
dipole .... 10, 87, 90, 96, 97-Ch. VIII - 100, 103, 108, 109, 124, 135, 199, 204, 214
donor 131
Doppler shift 50, 51, 63, 155, 202
Einstein .... 15, 18, 26, 29, 31, 33, 36-Ch. II - 38, 65, 69, 81, 82, 87, 89, 93, 98, 100, 109, 113, 139-Ch. XI - 141, 157, 183, 198, 205, 208, 241
electrical conduction 10, 17, 77, 127, 129, 138, 142, 207
electrical force equation 114, 219, 220, 229
electromagnetic force 81, 82, 139
electromagnetic induction 135
electron cloud 127
electron gas 128

electron orbit . 76, 96, 116, 117, 123, 128, 141-Ch. XI - 143, 158, 161, 162, 165, 176, 180, 181, 205, 206, 208-Ch. XIX - 210
electroweak unification 83
energy bands 128
ether 2, 25, 85, 98, 109
Faraday 22, 33
field ...... 2, 10, 15-- 17 -, 19, 22, 23, 37, 46, 55, 58, 69, 75, 81, 82, 85, 87, 89- Ch. VI - 91, 93, 95, 96, 97-Ch. VIII - 101, 103-Ch. VIII - 105, 107-Ch. VIII - 110, 112, 113, 127, 128, 132, 134-Ch. IX - 138, 139-Ch. XI - 142, 144, 154, 161, 173, 179, 181, 186, 191- Ch. XVII - 194, 197-Ch. XIX - 199, 204, 206-Ch. XIX - 208, 214, 220, 237
Fitzgerald 26
flat universe 58
force field 81, 87, 90, 93, 96, 110, 113, 199
Fourier analysis 41, 207
Galileo 21, 28, 29
Gamow 52, 54
gauge symmetry 83
gauge theory 82, 83
Gauss 22
global symmetry 83, 139
Goldstein 23
gravitino 84
gravity .. 1, 3, 4, 10, 13, 15-- 17 -, 21, 22, 62, 65, 67, 70, 81-Ch. VI - 84, 87, 89, 96, 110, 111-Ch. IX - 113, 123, 139-Ch. XI - 141, 143, 144, 145, 147, 149, 151-Ch. XII - 153, 188, 197, 198, 200, 201, 204, 205, 207-Ch. XIX - 210, 213, 214
Gribbin 52, 241
GUT 82, 84, 198
Hawkings 64
helium .... 54, 57, 62, 74-Ch. V - 76, 78, 79
Henbest 36, 241
Hertz 23, 42
Hill's equation 159
Hubble .... 18, 36, 50, 51, 54, 58-Ch. IV - 60, 64, 146, 188, 202
hydrogen atom . 10, 27, 44, 46, 52, 68-Ch. V - 70, 72, 73, 75, 77, 80, 86, 87, 96, 97, 99, 104, 107-Ch. VIII - 109, 111, 112, 116, 124, 125, 127, 141, 143, 144, 158, 159, 161, 173, 176, 188, 195, 199, 204, 213, 214, 219, 220, 229, 230
hydrogen engine 188
inductance .... 98, 132-Ch. IX - 135, 137, 143, 179-Ch. XV - 181
infinity 32, 33, 64

inflation 201
inner space 9, 31, 52, 67, 69, 74, 125, 211
Kepler 21
Law of Conservation of Energy 35, 155
Lorentz 26, 29, 140, 183, 198, 210, 239
Lorentz transformation 29, 183, 198, 239
magnetic flux .... 22, 135-Ch. IX - 137
magnetism .... 10, 17, 82, 83, 127, 132, 135-Ch. IX - 138, 140, 207
mapping .... 65, 163-Ch. XIV - 165, 211
Marconi 23
Mathieu's equation 206, 208
matter . 13, 16, 19, 22, 23, 26, 27, 29, 33-Ch. II - 37, 42, 44, 46-Ch. III - 48, 52,
54, 55, 62-Ch. IV - 64, 66, 67-Ch. V - 69, 72, 77, 82, 85, 88, 89,
93, 139-Ch. XI - 141, 145-Ch. XII - 147, 183, 191, 196, 198-
Ch. XIX - 200, 202, 203, 205
Maxwell 22, 29, 89
measurement problem 65, 127, 183, 184, 186, 188, 201, 203, 204, 210, 211
Michelson 25, 85
Milgrom 147
Mills and Yang 139
model ... 9, 16, 23-Ch. I - 29, 31, 44, 46-Ch. III - 48, 54, 61, 66, 67-Ch. V - 69,
71, 84, 85, 87, 90, 95, 96, 97, 99, 104, 108, 111-Ch. IX - 114,
123, 124, 144, 147, 154, 156, 158, 161, 165, 175, 178, 183, 188,
203, 213, 219, 220, 229, 230
Morley 25, 85
Newton 21, 22, 36, 89, 113, 114, 120, 122, 147, 217, 227, 228, 236
NMR 75
nonlinear analysis 95, 158, 162, 165, 206, 241, 242
nonlinear effect 95, 96, 97, 208
nonlinearity 95, 114, 116
permeability 98, 133, 134, 137, 179, 237
phlogiston 22
photon . 26, 28, 36, 69, 71, 76-Ch. V - 78, 99, 139, 143, 161, 175, 182, 184, 185,
201, 203, 206, 214
Planck .... 25-Ch. I - 28, 31, 33, 34, 69, 70, 156, 158
Plato 16, 20
plum pudding 23
Ptolemy 20, 29
QCD 82, 83
QED 82, 83
quantum theory . 25, 27-Ch. I - 29, 33-Ch. II - 35, 51, 52, 71, 72, 127, 128, 132,
161, 206, 207, 242
quarks 79, 83, 84, 86, 139, 194, 199, 205

radiation   17, 18, 23-Ch. I - 28, 31, 32, 34-Ch. II - 36, 43, 47, 51, 54, 63, 69, 71, 72, 76-Ch. V - 78, 100, 104, 107-Ch. VIII - 109, 126, 128, 137, 138, 142-Ch. XI - 144, 155-Ch. XIII - 159, 161, 162, 163-Ch. XIV - 165, 173, 176, 177, 184, 186-Ch. XVI - 188, 198, 200, 201, 203, 204, 207-Ch. XIX - 210
radiation curve 24, 25
radium 23, 156, 157
Richardson 131
rotating field .... 2, 96, 99-Ch. VIII - 101, 103, 104, 108, 186, 194, 204
Rutherford .... 26-Ch. I - 28, 44, 46, 47, 68, 87, 96, 97, 99, 104, 113, 114, 123, 144, 213, 219, 229, 230
Schrodinger 77
semiconductor 47, 77, 127, 131, 132
sine wave 40, 41, 166
singularity 32, 33, 125
Socrates 16, 20
spectra 10, 155, 157, 162, 164, 173, 176, 177, 208
spectrum ... 17, 25, 27, 32, 41-Ch. III - 43, 51, 69-Ch. V - 71, 76, 78, 128, 146, 155-Ch. XIII - 159, 161, 162, 164, 165, 176, 177, 208
speed of light .. 11, 17, 23, 25, 29, 34, 36, 37, 46, 55, 57, 58, 60-Ch. IV - 62, 64, 65, 97-Ch. VIII - 101, 104, 105, 107-Ch. VIII - 109, 137, 139, 142, 143, 179-Ch. XV - 182, 183-Ch. XVI - 189, 200, 202-Ch. XIX - 204, 206, 207, 210, 211, 214
speed of the electron 11, 34, 179, 181, 182, 186, 209, 210
Stahl 22
Stark 173, 176
Stark effect 173, 176
strong force 81, 82, 86, 141
supergravity 84
superparticle 84
supersymmetry 84
superunification 84
Theory of Relativity ... 17, 26, 29, 37, 52, 57, 58, 61, 64, 65, 67, 69, 82, 97, 100, 108, 139, 140, 183-Ch. XVI - 185, 188, 198, 204, 210, 239
Thomson 23
unified field theory ...... 15, 17, 81, 87, 137, 139-Ch. XI - 142, 191, 197, 198
universal force .... 9, 81, 82, 85-Ch. VI - 87, 91, 198
Volta 22
wave .. 11, 32, 40-Ch. III - 42, 68, 69, 77, 86, 99, 100, 107, 137, 140, 143, 163, 166, 179, 186, 191-Ch. XVII - 194, 206, 214, 237
weak force 62, 82, 83, 88, 141, 197
Wien .... 23-Ch. I - 25, 156